Oil Money and the World Economy

WESTVIEW SPECIAL STUDIES IN INTERNATIONAL ECONOMICS

Yoon S. Park, *Oil Money and the World Economy*
Karl P. Sauvant and Farid G. Lavipour (eds.), *Controlling Multinational Enterprises*

The whole gamut of national and international economic problems resulting from the oil crisis has proved difficult to manage, not only because of the size and suddenness of the oil-price increases but also because both scholars and government leaders suddenly found the world economy faced with a set of problems that had not been sufficiently explored in the postwar period. Conceived as an effort to put the oil crisis in perspective, this book examines how the crisis came about and how it might evolve in the future, then evaluates its serious effects on the world economy in the key areas of real income, inflation, standard of living, balance of payments, capital flows, international monetary reform, banking and finance.

Dr. Yoon S. Park is Senior Economist in the Treasurer's Department of the World Bank and a Professorial Lecturer in international finance at Georgetown University. In addition to various journal articles in the field of international economics and finance, Dr. Park's previous publications include *The Eurobond Market* (Praeger Publishers, 1974) and two Princeton University international finance essays, *The Link between Special Drawing Rights and Development Finance* (No. 100, September 1973) and *Artificial Currency Units: The Formation of Functional Currency Areas* (No. 114, March 1976). In addition to M.A. and M.B.A. degrees, he has received two doctorates: a D.B.A. in international finance from Harvard University's Graduate School of Business Administration and a Ph.D. in economics from The George Washington University.

Yoon S. Park
International Bank for Reconstruction and Development
and Georgetown University

OIL MONEY AND THE WORLD ECONOMY

With the collaboration of
JOSEPH ASCHHEIM, The George Washington University

Westview Press
Boulder, Colorado
WESTVIEW SPECIAL STUDIES IN INTERNATIONAL ECONOMICS

Published 1976 in the United States of America by

Westview Press, Inc.
1898 Flatiron Court
Boulder, Colorado 80301
Frederick A. Praeger, Publisher and Editorial Director

Library of Congress Cataloging in Publication Data

Park, Yoon S. 1935-
Oil money and the world economy.

Bibliography: p.
Includes index.
1. Petroleum industry and trade—Finance. 2. International finance. 3. Economic history—1945- I. Title.
HD9560.5.P25 338.2'3 75-40467
ISBN 0-89158-018-2

Printed in the United States of America.

Contents

Page

List of Tables and Figures

Figure

Preface

Since early 1973, the average price of crude oil from OPEC countries has more than quadrupled, causing the most serious disruption of the world economy since World War II. Various national and international economic problems resulting from the oil crisis have proved difficult to manage, not only because of the size and suddenness of the oil-price increases but also because they have been imperfectly understood by economists and policy makers alike. When the oil crisis broke out, both scholars and government leaders suddenly found the world economy faced with a set of problems that had not been sufficiently explored in the postwar period. While there has been no dearth of journalistic coverage of the oil crisis, most news stories merely documented tales of "gasless" Sundays, new restrictions on heating and outdoor lighting, and the comings and goings of experts attending endless energy conferences.

In the ensuing years, the number of oil crisis studies has steadily increased. Many of the early papers were hurriedly prepared, resulting in Cassandra-like predictions of world economic disaster. Some studies warned of the impending doom of major oil-dependent industrial countries such as Japan and France. Some foresaw the day when most of the world's financial wealth would be in the hands of several OPEC countries, and others drew grim portraits of the Western World overshadowed by the newly emerging economic powers of the Middle East. Since then, many calm and detached voices have been heard; still, there is a clear need for additional cool and *specific* objective analysis.

Conceived as an effort to put the oil crisis in perspective, this book first examines how the crisis came about and how it might evolve in the future. Its concern then shifts to the evaluation of such specific, major economic issues as the quadrupling of oil prices in the period 1973-74, which had serious effects on the world economy in the key areas of real income, inflation, standard of living, balance of payments, capital flows, international monetary reform, banking and finance.

In summary, this book attempts to address, in an evenhanded and consistent way, all the important economic questions raised by the oil crisis. Written not only for readers from the academic community but also for those from government and business sectors, it has tried to avoid complex mathematical expositions as much as possible. A general knowledge of the fundamentals of economics will suffice for comprehension. (The single exception may be the Technical Note to Chapter 4, which many readers may find convenient to ignore.)

The study could not have been completed without the help and encouragement of many people. Though space limitations prohibit an individual thank you to each of them, I wish to acknowledge especial gratitude to the following: Ching-Yao Hsieh and Robert Dunn, Jr., both economics professors at The George Washington University; Samuel Pizer of the U.S. Federal Reserve; Norman Klath of Morgan Guaranty Trust Company; Fred Klopstock of the Federal Reserve Bank of New York; and Pedro-Pablo Kuczynski of the International Finance Corporation. A very substantial debt is owed Professor Joseph Aschheim of The George Washington University, with whom I coauthored Chapter 8 (an earlier version appeared in the November 1975 issue of *Economia Internazionale*). I am also grateful to Messrs. Eugene H. Rotberg and Hans C. Hittmair, respectively Treasurer and Deputy Treasurer of the World Bank, for their continuous help and encouragement in all my research.

The views presented in this book are strictly my own and should in no way be ascribed to the World Bank or its management. Any errors or mistakes are my own responsibility.

Yoon S. Park
Washington, D.C.
March 1976

Chapter 1 Introduction

One of the major questions currently facing the world economy is whether the oil-importing countries and the international monetary system can adjust to the shock of the quadrupling of oil prices in 1973 and 1974 without serious economic disruptions. The growing burden of oil payments is being felt by most oil-importing countries in terms of inflation, balance-of-payments adversity, and detriment to the standard of living. International financial markets such as the Eurocurrency and Eurobond markets have been undergoing severe stress in seeking to cope with the recycling of oil money, and foreign-exchange markets are extremely uneasy because of the serious implications of oil payments for the currencies of oil-importing countries. The international banking community is experiencing a difficult adjustment period in intermediating huge oil-money deposits. Last but not least, international monetary reform has been set back seriously by the monetary and financial turmoil caused by oil-money flows. All these developments have created new issues and new problems in analyzing the world economy. It is not even certain whether all the main aspects of the new economic situation are sufficiently understood by either economists or policy makers.

The various economic problems mentioned above stem both from the oil-price upsurge itself and from the unusual suddenness with which it occurred. Thus, OPEC oil revenues have risen from $32 billion in 1973 to about $115 billion in 1974.[1] The combined current-account surplus of OPEC countries amounted to about $55 billion in 1974, and may be almost as much in 1975.

Estimates of the total reserve accumulation by OPEC countries range from $180 billion to $450 billion by 1980 in current dollars.

The sharp oil-price increase has created more than just economic problems for oil-importing countries. The international oil crisis of 1973 and 1974 also became a highly controversial political issue when members of the Organization of Arab Petroleum Exporting Countries (OAPEC) imposed a temporary embargo on petroleum shipments to many Western industrial countries for political reasons, concurrent with the quadrupling of oil prices by members of the Organization of Petroleum Exporting Countries (OPEC). Suddenly, oil-importing countries awoke to their heavy and increasing dependence on a handful of oil-exporting countries for a vital part of their energy needs. Oil-importing countries were caught by surprise in 1973, when the world petroleum industry underwent a revolution in which strategic crude-oil pricing and production decisions were taken over by the OPEC countries from the multinational oil companies. The members of OPEC then unilaterally quadrupled the price of crude oil and cut back oil production for political as well as economic purposes. This "revolution" in the international oil industry fundamentally changed the economics and the politics of oil in the world, and touched off an intense public debate over national energy policies in oil-deficient countries.

The full dimensions and ultimate consequences of this "revolution" will take years to unfold. What is needed in the meantime, however, is a rational evaluation of its various economic implications. The purpose of this study is to provide such an economic evaluation. The quintupling of oil prices in the past couple of years has had serious ramifications for the world economy. Since these ramifications have not been studied in a comprehensive manner, there is a general uncertainty about how to handle the critical economic issues raised by the oil crisis. This is exemplified by some of the conflicting statements made by public officials in the course of developing national policies to cope with the oil crisis. Some of these remarks are obviously political in nature, but others result from a genuine misunderstanding of the economic issues involved. Since the whole problem is rather complex, a piecemeal approach to its analysis will not be very useful. We can reach a workable understanding of the issues only through a comprehensive evaluation of all the economic factors related to the oil-price increase.

Accordingly, in this book we will attempt to study all the major aspects of the world economy which have been affected by the oil-price upsurge. The hypothesis of this study is that the international economic equilibrium has been disturbed by the sharp oil-price hikes in such a way that its adjustment process is not amenable to the typical macroeconomic demand-management policies implied in the Keynesian model. The internal balance of oil-importing countries has been disturbed by two opposite trends resulting from the oil crisis: on the one hand, a deflationary bias of the economy due to the "excise tax" effect of the oil-price increase on aggregate consumption, as well as to the decline in real income, and, on the other hand, an inflationary bias of the economy due both to the sharp price increase of oil products and oil substitutes, and the "ratchet effect" on prices of currency depreciation in oil-importing countries. At the same time, the external balance of oil-importing countries has been disturbed by massive oil payments. But the conventional remedies for an individual oil-importing country, such as currency devaluation or deflationary domestic economic policies, may not work for the oil-importing countries as a group because of the limited absorptive capacity of the oil-exporting countries, and because oil payments are made in the currencies of certain oil-importing countries rather than in the currencies of OPEC countries.

Consequently, the world economy finds itself in a new dilemma. The internal balances of oil-importing countries may not be restored by macroeconomic policies concerned with the aggregate demand, while their external balances cannot be achieved through conventional adjustment processes. In seeking to resolve this dilemma, this study develops (in Chapter 4) a new adjustment model, which incorporates the time element as a way to balance internal economic growth against external equilibrium.

In other words, our new model considers adjustment toward internal and external balances as a dynamic process involving a trade-off between internal economic growth and external balance.

Both in testing the hypothesis of this study and in developing the new adjustment model, we rely more on deductive reasoning, utilizing the analytical tools of economic theory, than on statistical inference. This methodological choice is partly intentional and partly imposed by circumstance. The current oil crisis came into existence only a couple of years ago. Consequently, we do not have enough serial data to run a statistical test of the economic variables which have been changed by the oil-price increase. More importantly, our basic concern here is to make a critical theoretical analysis of those economic problems that have been identified with the oil-price upsurge. Since the conceptual model used in this study derives from the theory of international economics, it is hoped that, when sufficient data become available in the course of time, the statistical validity of our model may be fully tested. For the time being, we have to be satisfied with using only estimated parameters in our calculations of the magnitudes of key dependent variables in the model.

The main body of this study is composed of eight chapters, each dealing with a key problem area connected with the oil-price hike. Chapter 2 traces the evolution of the international oil industry up to the present, emphasizing those factors that have contributed to the "revolution" of 1973 in world oil politics. Particularly, it studies how OPEC came into being and how it has evolved from an essentially defensive organization to an offensive one. Chapter 3 discusses the important issue of oil pricing, which had been the main point of conflict between oil companies and oil-exporting countries until 1973, when the latter took over the oil-pricing power completely. Oil pricing has always been regarded as a rather complicated subject, and rightly so, but there can be no real understanding of the international oil industry without first comprehending the many issues concerned with oil prices.

Chapters 2 and 3 prepare us for studying the substantive issues connected with the economic disruption caused by the oil-price increase. Chapter 4 considers the transformation that has taken place in the world balance-of-payments picture due to increased oil payments, and studies several ways to cope with this problem. Chapter 5 shows how the increase in oil prices has affected the real income of oil-importing countries. Chapter 6 studies the effect of oil-money flows on international financial markets such as the Eurocurrency and Eurobond markets. Chapter 7 measures the change in international liquidity due to increased oil payments and its impact on the worldwide inflationary trend. Chapter 8 surveys various recycling proposals designed to deal with the oil-financing problem and analyzes the theoretical implications of the transfer problem inherent in the current oil crisis. Finally, Chapter 9 considers the special problems facing non-oil developing countries as the result of quadrupled oil prices and the subsequent dislocations in the world economy. Chapter 10 summarizes our principal findings and suggests areas for further research.

NOTES

[1]*Oil-Related Payments Imbalances,* Bank for International Settlements (Basle, Switzerland, March 11, 1975). According to the estimate by the U.S. Treasury, the OPEC "government take" of oil revenues was $25 billion in 1973 and $95 billion in 1974. See U.S. Treasury Department, *The Absorptive Capacity of the OPEC Countries* Washington, D.C.: September 5, 1975).

Chapter 2
Geopolitics of Oil Versus Economics of Oil

Man's progress from a primitive stage to the complex industrialized society of today has required an increasing use of non-human energy. To an ever-increasing extent, energy has been used to move machines and vehicles of transportation and to provide heat and light for daily living. In fact, the development of human civilization has been shown to be closely tied to the degree of energy use per man. According to a study,[1] an average person in the United States currently consumes about one million BTU (1 BTU = 0.25 Kcal) of energy per day, which is a hundred times the amount used by "Primitive Man" in 100,000 BC. Even today, per capita energy consumption differs greatly from country to country. The average per capita consumption in the United States is about five times that in the world as a whole, twice that in Europe, ten times that in India, and a hundred times that in rural Africa. Table 2.1 (p. 8) shows percentages of world energy consumption by various regions.

In the contemporary world, most energy is supplied from coal, crude oil, natural gas, hydro sources, and nuclear fuels. Cheaper energy increases the real income of a nation by reducing the costs per unit of output. It has been estimated that, until the Industrial Revolution, some 80-85 percent of the total energy used by human societies was provided by plants, animals and men. With the Industrial Revolution, however, the coal age arrived; up to 50 years ago coal provided 85 percent or more of world energy requirements.

Table 2.1
Energy Consumption and Population, by Region (Percentages)

Region	Energy Consumption			Population			Average Annual Rates of Growth, 1960-1970	
	1950	1960	1970	1950	1960	1970	Energy Consumption	Energy Consumption per capita
Developed Countries								
United States	44.5	36.1	31.4	6.1	6.0	5.7	4.2	2.9
Canada	3.5	3.1	3.3	0.5	0.6	0.6	6.1	4.2
Western Europe	22.8	21.0	22.3	12.1	10.9	9.9	6.3	5.3
Oceania	1.2	1.1	1.1	0.5	0.5	0.5	5.8	3.5
Japan	2.3	3.0	5.2	3.3	3.1	2.9	11.9	10.7
U.S.S.R.	11.0	14.4	14.9	7.2	7.2	6.7	6.0	4.7
Eastern Europe	5.7	6.5	6.0	3.6	3.3	2.9	4.7	4.0
Developing Countries								
Latin America	3.1	4.0	4.3	6.5	7.1	7.8	6.3	3.4
Africa[a]	1.7	1.7	1.7	8.7	9.2	9.7	5.4	3.0
Asia (Communist)	1.6	5.3	5.3	22.8	22.7	22.0	5.6	3.9
Asia (Other)	2.7	3.7	4.5	28.9	29.4	31.3	7.7	5.0
World	100.0	100.0	100.0	100.0	100.0	100.0	5.6	3.6

* South Africa is included.

Source: Joel Darmstadter and Sam H. Schurr, "The World Energy Outlook to the mid-1980s: The Effect of an Alternative Supply Path in the United States," paper presented to Royal Society meeting "Energy in the 1980s," London, November 15, 1973, Table II, and James W. Howe, *The U.S. and the Developing World: Agenda for Action 1974* (New York: Praeger Publishers, 1974), Table C-1.

DEVELOPMENT OF THE WORLD OIL INDUSTRY

Compared to coal, petroleum was a latecomer, and assumed importance as a source of energy only after the First World War. Commercial production of petroleum first started in Rumania, the United States and elsewhere in the 1850s. Total annual world output reached 100 million tons (766 million barrels) by 1921, stimulated by the increasing use of the internal combustion engine for transportation. Developments in the world energy mix since 1925 are shown in Table 2.2. The table indicates that there has been an increasingly rapid shift in energy sources from coal to oil and gas. From 1925 to 1970, total energy consumption rose at a compound annual rate of about 3 percent. A breakdown into sub-periods shows a recent acceleration of energy consumption; the growth rate of energy consumption for the 1925-1950 period was less than 2 percent annually, while the growth rate for the years from 1950 through 1970 was about 5 percent, two and a half times that of the earlier period.

Table 2.2
World Consumption of Primary Energy: 1925-1970

	1925	1950	1970
Total primary energy in MToe[1]	1,200	1,800	5,100
	percent distribution		
Coal	82%	60%	34%
Oil	13	25	40
Natural gas	3	10	20
Hydro	2	5	6
Total	100%	100%	100%

[1]Energy is expressed in millions of tons of oil equivalent (MToe).

Source: International Atomic Energy Agency Bulletin, Vol. 15, No. 5, 1973.

The substitution of petroleum for coal as an energy source increased even more from 1960 through 1973. The average annual growth rate of consumption during this period was only 0.3 percent for coal, while it was about 7 percent for oil and gas. This development was especially pronounced in Western Europe and Japan, where oil and gas consumption grew at an annual rate of 10 to 20 percent. In contrast, the use of coal was actually declining in Europe and increasing only slowly in Japan. As a result of such dramatic growth, the oil industry has become the world's leading industry in terms of assets over the past decade. Due to the geographical separation between regions of mjor oil production and regions of high oil consumption, oil is also first in importance in its contribution to the world's tonnage of shipping, as well as to the value of international trade.

Especially since the end of the Second World War, growth in oil has exceeded growth in all other large-scale industries in terms of the scale of operations. To meet the rapid growth in demand, oil production has accelerated steadily. In 1950, twice as much crude oil was produced as had been in 1945. Ten years later, in 1960, annual crude-oil production had doubled again to 1,000 million tons, and by 1968, production had reached 2,000 million tons. But generalizations about the *world* oil industry may be somewhat misleading, because the oil industry is not internationally homogeneous. There is a clear disjunction between the oil industry of Communist nations and that of the rest of the world, and within the latter category a split has emerged in the post-war period between the oil industry of North America and that of the remaining non-Communist World.

Up to 1945, the United States was the leading exporter of petroleum products, but it soon lost that position because its domestic demand for oil overtook its domestic crude production. By 1948, the oil imports of the United States exceeded its oil exports and so, for the first time in the history of the oil industry, the United States became an oil-deficit nation. Thereafter, U.S. oil imports continued to grow, and with them rose the concern for "security of supply." This concern has led to an increasingly autarchic oil policy on the part of the United States. In this respect, U.S. oil policy has been similar to that of the Soviet Union, which has protected its domestic oil industry through the exclusion of imports.

Outside the United States and the Communist-bloc countries, the oil industry has been dominated by the seven largest oil companies, known as the "international majors." According to an expert in this field,[2] as of the early 1970s these seven oil companies were responsible for about 80 percent of all oil production in the world outside the Communist countries and North America. In the same areas they owned or controlled over 70 percent of the total refining capacity, and they operated, either directly or indirectly, well over 50 percent of the tonnage of internationally operating tankers. A list of the seven international majors follows:

Seven "International Majors"	**Short Name**
Exxon, formerly Standard Oil Company (New Jersey)	"Exxon"
Mobil, formerly Socony-Vacuum Oil Company	"Mobil"
Gulf Oil Corporation	"Gulf"
Texaco, formerly The Texas Company	"Texaco"
Standard Oil Company of California	"Socal"
British Petroleum Company, Ltd.	"B.P."
Royal Dutch Petroleum Company and Shell Transport and Trading	"Shell"

The global operations of the international majors were aided and protected by their respective governments, which, as early as the end of the 19th century, became fully conscious of the growing economic, military and political importance of petroleum. Lord Curzon, a member of the British War Cabinet, said at the close of World War I: "The Allies floated to victory on a wave of oil."[3] Western industrial countries, therefore, competed with one another in their efforts to gain access to the world's petroleum deposits. In order to secure exclusive rights to foreign oil fields, oil companies first negotiated the necessary legal grants from the rulers of the Near and Middle East countries through the "concession," which was to become a notorious institution of the major oil companies. The terms of a concession granted to oil companies the exclusive rights of research, extraction, export and sale of petroleum, as well as immunity from taxes and customs duties, in return for the payment of fixed royalties to the foreign government.

The concessions, which had been granted at various times between 1901 and 1935, stipulated a fixed royalty of 20-25 cents per barrel of crude oil produced. On an average, the Middle Eastern governments received 21 cents a barrel, a figure close to the Venezuelan one of 23 cents.[4] These arrangements may have been fair enough at this early time, when the demand for oil was growing slowly and crude-oil prices were fluctuating or sagging, when the prospects for striking oil had not been fully tested, and when huge capital investments had to be made in far-away foreign regions. However, the sharp rise in prices, including that of oil, during and after the Second World War, reduced both the purchasing power of the fixed royalty and the local governments' share of the value of each barrel of oil—in Iran, for example, the percentage dropped from one-eighth before the War to one-sixteenth in 1947.

Apart from the economic provisions, the concessions granted to foreign oil companies excessive territorial rights, leading the companies to appropriate to themselves a quasi-colonial authority. The area as well as the duration of the concessions was excessive. For example, the D'Arcy Concession, the first oil concession (granted in 1901), covered the entire territory of Iran, with the exception of only five provinces. Concessions at Kuwait and Qatar comprised the whole of the territories, and that in Iraq covered the entire country to the east of the Tigris, with the exception of a couple of provinces. In addition, the duration of the concessions was relatively long, varying from sixty to seventy-five years, and even this period was later extended. For instance, the D'Arcy Concession was extended in 1933 from sixty to ninety-two years, and the concessions at Kuwait and Bahrein were later extended to ninety-two and ninety years, respectively.

The growing discontent of oil-producing countries with the concessions first led Venezuela to impose taxes on oil in 1943. The tax rate was gradually increased, and by 1948 the government had secured half the net income earned in oil production by the oil companies. This was the beginning of the so-called 50-50 arrangement, which raised payments to the Venezuelan government to about 75 cents per barrel of oil produced. This development in Venezuela constituted a most important occurrence in the history of relations between oil-producing countries and oil companies. The concessionaire's monopoly was broken, and in the place of long-term concessions, the Venezuelan government adopted the system of granting short-term licenses for prospecting over a limited area.

In 1949, the Venezuelan government sent a mission to the oil-producing countries of the Middle East in order to draw their attention to the advantages of the 50-50 arrangement instituted in Venezuela. The Venezuelan mission was not entirely generated by the brotherly altruism of a fellow oil-producing country; only if other major oil-producing countries acted in unison to assert their rights could Venezuela maintain its competitive position in world oil production. In any case, the advantages of the 50-50 arrangement in place of low fixed royalties were obvious, and other oil-producing countries wanted to follow suit. In the Middle East, the first 50-50 arrangement was concluded between Saudi Arabia and the oil companies at the end of 1950, and it was followed a few months later by similar agreements in Kuwait and Iraq.

In the meantime, growing dissatisfaction in Iran with the D'Arcy Concession led to the political crisis of 1951, in which the petroleum industry was nationalized by the Iranian government, nullifying the D'Arcy Concession. The Iranian crisis and its outcome were a grave warning to all oil companies, which were forced to reassess their strong resistance to the 50-50 arrangement. By the end of 1952, therefore, the 50-50 profit-sharing system had been extended to the remaining oil-producing countries, and in Iran, a compromise was finally reached in 1954 to adopt the 50-50 system instead of outright nationalization.

Thus, by the early 1950s the long era of concessions had been replaced by the new era of profit sharing.

The adoption of profit sharing coincided with an emerging challenge to the international majors by other oil companies. During the ten years following the Second World War, the international oil industry was dominated by the seven majors. In the United States, however, there were a number of oil companies, known as the "independents," that as a group produced more oil than the majors in North America. In 1952, for example, the combined production of the majors in North America amounted to 764 million barrels, while that of the independents in the same region was 1,587 million barrels, more than twice the amount produced by the majors.

During the 1950s, the independents of the United States and oil companies of other Western industrial countries such as Italy and France began to enter an international oil scene in which conflicts between the oil-producing countries and the international majors were gathering intensity. The independents found favor with the producing countries, because they showed a willingness to accept the stiffer conditions now asked for by the producing countries, in contrast to the reluctance of the entrenched majors to give in to the demands of the oil-producing countries by revising the existing contracts.

ESTABLISHMENT OF OPEC

The advent of the profit-sharing arrangement introduced a new element into the relationship between the producing countries and the oil companies. During the era of concessions, the oil revenues of producing countries consisted only of the fixed royalties per barrel of oil produced; hence their revenues varied solely with the volume of output. With the profit-sharing regime, however, government revenues depended on oil price as well, because the producing countries would receive half the difference between the posted selling price of oil and its cost of production. However, oil prices had traditionally been set by oil companies that controlled both production and marketing in the Western world. As long as the producers of crude oil were also the buyers of that same oil, the price of crude could be arbitrarily set. The international majors, controlling the production, refining, and marketing of oil, could therefore influence the amount of oil revenues to the oil-producing countries, even in a 50-50 arrangement, by changing the selling price of crude. The selling price of crude oil established by the majors was nothing more than an internal accounting price from their production subsidiaries to their refining subsidiaries. The majors usually changed the "selling" price without consultations with the producing countries. The latter came to possess a keen interest in oil prices, however, because the prices determined oil revenues to them under the 50-50 arrangement. The Organization of Petroleum Exporting Countries (OPEC) was established by the major oil-producing countries in 1960 to prevent unilateral price reductions by the oil companies, and, if possible, to raise oil prices and thus increase government oil revenues.

Earlier attempts to coordinate the producing countries' policies towards oil companies had begun in the mid-1940s, first among Arab oil countries within the framework of the Arab League, which was established in 1945. In 1947, diplomatic contacts were established in Washington between Venezuela and Iran for the purpose of coordinating their oil policies. Iran was at that time commencing its negotiations with the oil companies to revise the D'Arcy Concession.[5] Further progress toward the goal of better coordination was made in 1949, when Venezuela sent a special three-man delegation to visit Iran, Iraq, Kuwait and Saudi Arabia, and persuaded these countries to adopt a profit-sharing regime in place of concessions. On their part, the Middle East oil countries had also recognized the need to unite all oil-producing countries for better protection of their interests against the giant oil companies backed by their powerful governments. From its beginning, the Arab League had entertained the idea of an Arab-wide petroleum association.[6]

Interest among producing countries in a close coordination of their oil policies increased sharply after the widespread adoption of the 50-50 arrangement. In 1953, Iraq and Saudi Arabia signed an agreement of cooperation on oil matters, which was the first agreement between oil-producing countries calling for cooperation at the government level in the petroleum field. The agreement provided for the exchange of oil information and the holding of periodic consultations on petroleum policies. However, in order for an oil producers' association to be effective, other non-Arab countries, especially Iran and Venezuela, had to be included. Serious thinking on the establishment of an association of oil-producing countries was delayed until 1959, when the first Arab Petroleum Congress met in Cairo in April, under the sponsorship of the Arab League. Representatives of Venezuela and Iran were invited to attend as observers.

It was during this Cairo conference that "the first seed of the creation of OPEC" was sown.[7] However, the immediate cause for the creation of OPEC was another unilateral reduction of the crude oil price in August 1960. As explained before, under the profit-sharing arrangement, a decline in the crude oil price means a decline in the government revenues of oil-producing countries. The price reduction of August 1960 caused Middle East oil countries a loss of about $93 million for 1960 alone.[8]

One month after the August price reduction, representatives of five oil-producing countries (Iran, Iraq, Kuwait, Saudi Arabia and Venezuela) met in Baghdad and decided, on September 14, 1960, to create OPEC. As an observer later noted, "The attitudes that were finally expressed in the formation of OPEC had begun to form years before, too. But it took the price cuts of August, 1960 . . . to crystallize them."[9] Therefore, the dissatisfaction of oil-producing countries over unilateral oil-price cuts by oil companies was the direct cause for the establishment of OPEC. What then were the reasons for the two oil-price reductions in 1959 and 1960, which immediately preceded the establishment of OPEC? In 1959, the already weak world crude-oil market was further weakened when the United States imposed mandatory import quotas on both crude oil and oil products, in order to reduce dependence on foreign oil for security reasons. This move effectively closed the U.S. market for all oil imports except those from Canada and Mexico. Oil from these countries was excluded from the restrictions on the grounds that no risk was involved to its importation, as this did not depend on ocean transportation.

The U.S. concern for "security of oil supply" had been played up during the early 1950s by both politicians and domestic oil interest, whose efforts were aided by the political difficulties engendered by the cold war and the Korean War. Consequently, the Eisenhower administration was forced to take action in 1954, first calling for voluntary restraints on oil imports. The larger companies heeded the warning, partly because they could market their overseas crude oil in foreign markets through their long-established international marketing organizations. However, the voluntary import restraints could not stop other oil companies, especially those U.S. independents that were now bringing their new overseas oil fields into production, from taking advantage of the voluntary nature of this program to undercut the larger companies.

As the voluntary program turned out to be largely ineffective, the U.S. government imposed a mandatory import quota system in 1959. The mandatory program had far-reaching effects on the U.S. and international oil markets. The U.S. import quotas were set roughly at the level of the percentage contribution of imports to the total U.S. oil consumption in 1959, which was approximately 12 percent. Since U.S. oil consumption was growing at 3 percent annually, the amount of oil imports could grow only at 3 percent per year, compared with a 15 percent annual rate of growth over the previous decade. At the same time, domestic oil interests were guaranteed about 88 percent of the existing U.S. demand, as well as 88 percent of all incremental demand. Many of the independent oil companies, which had invested heavily in the 1950s in oil explorations outside the

United States in the hope of selling their output in the U.S. market, were now effectively shut off. Consequently, they were forced to dump their foreign-produced oil in the international market, already largely preempted by the majors, at steep discounts.

The majors, with large reserves of low-cost Middle East crude, had to match the prices of the independents by granting comparable discounts. At the same time, a growing volume of Russian oil was reaching the European market. As the result of the widening gap between the posted prices and the market prices (to be explained in the next chapter), the profit margin of the international majors declined, since the 50-50 profit-sharing arrangement was based upon the posted price and not on the realized market price. In order to increase the profit margin, the oil companies cut the posted prices of Middle East oil by about 18 cents per barrel of crude oil in February 1959, without prior consultation with the producing countries. In August 1960, the oil companies again unilaterally cut the posted prices by an average of about 9 cents per barrel, ignoring the Arab warning in the spring of 1959 in the form of a resolution adopted by the first Arab Petroleum Congress calling on oil companies to consult the producing countries before making any price changes. These two successive cuts inflicted a loss of about 13.5 cents per barrel in the oil revenues of the producing countries, which meant a combined loss of about $231 million for Kuwait, Iran, Iraq and Saudi Arabia in 1960.[10]

In 1960, when OPEC was established, its five founding members held 67 percent of world oil reserves and accounted for 38 percent of world oil production and nearly 90 percent of internationally traded oil. Eight other countries subsequently joined OPEC: Qatar (1961), Indonesia and Libya (1962), Abu Dhabi (1967), Algeria (1969), Nigeria (1971), Ecuador and Gabon (1973). In 1974 Abu Dhabi's membership was passed to the United Arab Emirates, which it joined in 1971. The member countries of OPEC now account for 85 percent of the oil entering international trade. During the first ten years of its existence,

OPEC's main function was defensive—trying to prevent further erosion in the posted prices. Consequently, there has been no more reduction of the posted prices by oil companies from 1960 on, even though the market prices of oil eroded further during the 1960s. The other main achievement of OPEC during the 1960s was the agreement reached in 1964 between the OPEC countries and the oil companies to change the methods by which oil-company profits were calculated. The net effect of the change was to increase government revenues per barrel, initially by 4 cents in 1964, and by about 11 cents by 1970.

During the 1960s, another association of oil-exporting countries was established. In January 1968, Saudi Arabia, Kuwait, and Libya signed an agreement in Beirut creating the Organization of Arab Petroleum Exporting Countries (OAPEC). Since the membership of OPEC included countries with diverse attitudes in regard to particular political questions, Arab oil countries wanted a separate association, in order to find ways to utilize their petroleum as an effective instrument of their common political policy. The need for such an organization presented itself in connection with the Arab-Israeli conflict of 1967. Even though Arab oil countries wanted to suspend the export of their petroleum to countries that were presumed to be in sympathy with Israel, OPEC could not provide a forum for a discussion of this issue, since it was distinct from questions of petroleum policy relevant to all OPEC members. The impetus for the creation of OAPEC originated in these circumstances.[11]

By 1971 five other countries had joined OAPEC: Abu Dhabi, Algeria, Bahrein, Dubai, and Qatar. In December 1971, the membership rules were changed to allow all Arab oil-exporting countries to join OAPEC, regardless of whether oil provided "the principal and basic source of their national income." In 1972 Egypt, Iraq and Syria became OAPEC members. The statutes of OAPEC make it clear that its creation does not modify its members' obligations toward OPEC, and that OAPEC members are to respect OPEC resolutions even if they are not OPEC members.

Up to 1970, the world oil industry had been operating on the basis of a *modus vivendi* between oil companies and OPEC members. Even though the posted prices were not reduced by the oil companies due to the united front of the OPEC countries, the market prices of oil were largely determined by supply and demand, and only the oil companies controlled the amount of oil to be produced. As long as OPEC members did not possess the power to control and regulate the production volume of oil in their countries, they could not "dictate" oil prices unilaterally. Paradoxically, it was the earlier encouragement that the producing countries had given to the independents to participate in oil exploration and development that stood as the major obstacle to the achievement of production and price controls by the OPEC members. If the producing countries had had only the seven international majors to deal with, then some form of oil output regulation, acceptable to both sides, would probably have been achieved sooner within a framework of bilateral oligopoly. However, in the 1960s the majors were forced into fierce competition for markets both with the independents and among themselves, so that the prospect of a worldwide prorationing of oil production—whereby production could be tied to estimated demand at a given price—became more remote than it had been at the time of OPEC's establishment.

A NEW STAGE FOR THE WORLD OIL INDUSTRY

During the early 1970s, the international oil industry underwent revolutionary political and economic changes, due primarily to three factors: global price inflation in the late 1960s while oil prices remained almost constant; rising nationalism among oil-producing countries; and the Arab-Israeli conflict.[12] The first disturbance in the seeming equilibrium of the international oil market was triggered by Libya. By early 1970, petroleum consumption in Europe and North America had spurted upward much faster than expected. Since Libya is located close to Europe, oil production there was stepped up rapidly to meet European demand. Since the closing of the Suez Canal in mid-

1967 during the Six Day War, the increase in tanker rates had put Libyan oil in a strong competitive position. Finally, in May 1970, the Trans-Arabian Pipeline, which carried 500,000 barrels of crude oil a day from Saudi Arabia to the Mediterranean, was shut down due to an accident and remained closed for the rest of the year, thus increasing the importance of Libyan oil to Europe.

Early in 1970, the revolutionary government of Libya, which had overthrown the regime of King Idris in the preceding year, used its increased bargaining power to demand a large raise in oil revenues. After an arduous round of negotiations, the oil companies operating in Libya agreed, in September 1970, to raise the posted price by about 13 percent. Within several months this price increase spread to other oil-producing countries such as Iraq, Iran, Kuwait, Saudi Arabia and Nigeria.

This initial success encouraged OPEC to take a new stand. In December 1970, OPEC's Twenty-first Conference met in Caracas and called for immediate negotiation with the oil companies to attain the following objectives: to establish 55 percent as the minimum rate of taxation; to bring about general increase in all posted prices; to eliminate disparities in prices among the OPEC members; and to eliminate special discounts on sales of crude oil. The unprecedented solidarity of the OPEC members in pushing their new demands led many governments, especially the U.S. government, to give their oil companies antitrust clearance to enable them to negotiate as a group with the OPEC countries. The major oil-consuming countries had learned their lesson from Libya, whose government had forced its demands on the oil companies one by one, starting with the weakest. The consumer governments' representatives met in Paris to formulate a common policy in support of the oil companies in their negotiations with the OPEC countries.

Having thus closed their ranks and obtained assurances of diplomatic support from their respective governments, fifteen oil companies sent their representatives to meet with the OPEC representatives. Under the threat of unilateral legal measures to implement OPEC demands, however, the oil companies were forced to substantially meet these demands in the historic Teheran Agreement of February 1971, which was supposed to run for five years. In the Caracas Conference of December 1970, the OPEC members had also called for future adjustments of oil prices to reflect changes in the exchange rate of currencies. This led to the Geneva Agreement in January 1972, which raised the price of oil by 20 cents per barrel to offset the U.S. dollar devaluation of August 1971, and to a second agreement, in June 1973, to raise the price by 15 cents per barrel to offset the dollar devaluation of February 1973.

In September 1973, the OPEC members called for renegotiation of the 1971 Teheran Agreement, noting that the current rate of inflation far exceeded the 2.5 percent rate contemplated in the agreement. The oil companies proposed that, instead of renegotiating the whole Teheran package, an escalator clause be written into the agreement to cause oil prices to rise *pari passu* with general inflation. But the negotiations were never completed, and when a new Arab-Israeli War broke out in October, the OPEC members in the Persian Gulf *unilaterally* increased the posted price of crude oil by an unprecedented amount.

Thus, for the first time in the history of the oil industry, the pricing power was taken over completely by the oil-producing countries. The October 1973 increase was followed by another big increase in January 1974. Within two years, the posted price of crude oil had more than quadrupled. For example, the posted price of Saudi light crude increased from $2.59 per barrel in early 1973 to $11.65 per barrel in January 1974.[13] In 1973, the oil-producing countries completed the takeover of both crude-oil pricing and production decisions, which had traditionally been under the control of the oil companies. Moreover, many producing countries have progressively nationalized their oil industries or acquired majority ownership of oil reserves and producing facilities located within their boundaries. Consequently, the Western world is confronted with a new set of economic issues arising from the drastic changes in the oil industry. Over the past several years, the position of the oil companies has been reduced to that of agents for the production and marketing of oil, and substantive oil bargaining has become more and more a political matter between the governments of producing and consuming countries.[14] Considering the importance of oil for contemporary economic life, as well as the importance of the huge surplus oil revenues being accumulated by many OPEC members, we can understand the close interrelationship between the geopolitics of oil and the economics of oil.

NOTES

[1]*Energy Supply/Demand Outlook 1980-1985*, International Bank for Reconstruction and Development (Washington, D.C., July 18, 1974), p. 1.

[2]Peter R. Odell, *Oil and World Power: Background to the Oil Crisis*, 3rd ed. (Baltimore, Maryland: Penguin Books, 1974), p. 11.

[3]Neil H. Jacoby, *Multinational Oil* (New York: Macmillan Publishing Co., 1974), p. 27.

[4]Charles Issawi, *Oil, the Middle East and the World* (LaSalle, Illinois: The Library Press, 1972), p. 33.

[5]Zuhayr Mikdashi, *The Community of Oil Exporting Countries* (London: George Allen & Unwin Ltd., 1972), p. 23.

[6]Fuad Rouhani, *A History of OPEC* (New York: Praeger Publishers, 1971), p. 76.

[7]Perez Alfonzo, "The Organization of Petroleum Exporting Countries," *Monthly Bulletin*, No. 2, Ministry of Mines & Hydrocarbons, Caracas, 1966.

[8]Fuad Rouhani, *op. cit.*, p. 77.

[9]J.E. Hartshorn, *Politics and World Oil Economics* (New York: Praeger Publishers, 1962), p. 20.

[10]Abdul Amir Kubbath, *OPEC: Past and Present* (Vienna: Petro-Economic Research Center, 1974), p. 13.

[11]Fuad Rouhani, *op. cit.* , p. 161.

[12]Neil H. Jacoby, *op. cit.*, pp. 258-59.
[13]It was only $1.80 per barrel as recently as 1971.
[14]James Akins, "The Oil Crisis: This Time the Wolf Is Here," *Foreign Affairs*, April 1973, p. 478.

Chapter 3
Pricing Practices in the International Oil Industry

The price of oil, or more accurately the price of crude oil, has been one of the most emotional issues in the controversy between the oil-producing countries and the oil companies. We have seen in the previous chapter that the primary cause for the creation of OPEC was the controversy over crude-oil prices after the oil companies unilaterally cut the posted prices in 1959 and 1960. However, the oil-producing countries had not shown much interest in the issue of oil prices until the end of the 1940s, because their oil revenues, paid mostly in the form of royalties, were not based on the price of crude oil but on the level of crude-oil production, at so many cents per barrel of oil produced.

However, the situation changed completely with the widespread adoption of the 50-50 profit-sharing regime in the 1950s. As noted earlier, the profit-sharing regime originally started with the 1943 Venezuelan law concerning the relationship between Venezuela and the oil companies. Subsequent amendments to this law in 1945, 1947, and 1948 secured to the Venezuelan government at least half the oil companies' profits. The adoption of the 50-50 arrangement made it essential for a producing country to familiarize itself with the question of oil prices because government oil revenues were now tied to them. The difficulty, however, was that the "oil price" meant different things to different groups. The adoption of a particular pricing system would greatly influence the ultimate pattern of cost as well as revenues for the oil-producing countries, oil companies, and consumers.

DEFINITION OF CRUDE OIL PRICES

For crude oil, there are in general three different prices: the market price, the transfer price, and the posted price. In order to understand the meanings of these prices, it is necessary to see how the international oil industry is organized. In the Western world outside the United States, most of the crude-oil production is in the hands of the international majors, which have developed highly integrated operations, including exploration, production, transportation, refining, and marketing, through various affiliates. It is estimated by most observers that currently at least 85 percent of the crude oil traded internationally is not sold on the free market, but passes from one affiliate to another of a parent oil comany which has integrated oil operations. Therefore, prices for 85 percent of the internationally traded crude are "determined" by oil companies within the framework of their integrated worldwide operations and not by the competitive forces of free-market transactions.

This internally determined crude price is called the transfer price, which is a mere accounting concept, devoid of the economic function of equilibrating supply and demand. The specific level of these transfer prices is generally so determined as to achieve a maximization of the company's consolidated net profits, taking into consideration the different tax systems in the various countries where the company's affiliates are located. Like other multinational corporations, integrated oil companies "allocate overhead costs among their foreign branches, subsidiaries and affiliates, and adjust their transfer prices, in order to reduce their total tax outlays."[1]

In contrast, the market price is the actual crude-oil price realized under the competitive conditions of supply and demand in a free market. Since about 85 percent of the internationally traded crude never passes through the free market, the market price is relevant only for the remaining 15 percent or so. Although the free-market transactions thus account for only a small share of world crude movements, they have had a great effect on the development of the international oil industry. Without the free market, the international majors would have had a complete monopoly over the crude market and would not have been forced to cut the posted prices in 1959 and 1960, thus directly triggering the formation of OPEC. The world crude-oil market in the late 1950s and the 1960s was, in general, a buyer's market characterized by oversupply and fierce competition among crude sellers. The free-market transactions were made between independent refiners with no crude oil production of their own on the one hand, and independent oil producers with no downstream outlets, as well as the international majors, on the other. The independent producers, blocked from marketing their crude in the United States due to the U.S. import quota system, were forced to dispose of their crude to independent refiners. They could afford to price their oil competitively by taking advantage of low production costs and large profit margins. The international majors, on the other hand, met the competition of the independents by allowing similar price discounts, in order not to reduce their overall market share.

In practice, however, even the realized market price did not reflect the market value of crude oil as governed by supply and demand under open competition. The market prices being reported by trade publications, international bids, customs statistics, etc., were blurred by numerous non-price discounts, such as freight concessions, tied loans at favorable interest rates, generous terms of delivery and terms of payment, buy-back commitments, technical assistance, and favorable currency treatment.[2]

THE POSTED PRICE

To the oil-producing countries, the posted price is the most important price, since their revenues are based on it. As mentioned earlier, the 50-50 profit-sharing arrangement required a specific price (or tax reference price) for dividing the "profits." In general, the "profits" to be divided evenly between a producing country and oil companies under the 50-50 regime are the residual of the reference price after deducting the cost of production. However, the transfer price cannot serve as the reference price, since it is really nothing but an internal accounting price for integrated oil companies. Therefore, the tax reference price has been the posted price of crude in the free market. Historically, the posted price originated in the United States, where crude oil was sold openly and actively by numerous independent and major producers to equally numerous independent or integrated refiners. In other words, in contrast to international practice, bulk of crude in the United States was traded in a competitive market. Consequently, the market price of crude in the United States reflected actual supply and demand conditions very closely. The market prices so determined were posted (or announced) by refiners (buyers), thus defining the prices they were willing to pay for crude oils from any producer. Outside the United States, most crude production was under the control of a few highly integrated and diversified large-scale producers, and market prices were posted by the sellers rather than by the buyers. As mentioned earlier, however, there developed an oversupply of crude in the 1950s as a result of aggressive marketing by the independents, which were largely blocked from penetrating the U.S. market due to the U.S. import quota system. In the late 1950s, therefore, the posted prices gradually lost their significance as indicators for the real market prices for crude oil, since the realized market prices increasingly diverged from the posted prices. Confronted with strong reactions from the oil-producing countries, the international majors were reluctant to reduce the posted prices even though the real market prices declined. The main function of the posted prices, therefore, was to serve as tax reference prices for both the producing countries and the oil companies in calculating the government oil revenues under the profit-sharing arrangement. As a result, the posted price became most important to the producing countries; it was the unilateral lowering by oil companies of the posted prices which led to the establishment of OPEC. There are different posted prices for crudes, depending upon distances from the major markets, and technical factors such as the sulfur content and gravity of the oil. These technical aspects of crude price differentials have been extensively discussed in trade journals and books.[3]

The history of some of the posted prices for OPEC crude oils is shown on the following page.

DETERMINATION OF GOVERNMENT TAKE BY PRODUCING COUNTRIES

In terms of our study, it is instructive to see the way in which government revenues of oil-producing countries are calculated on the basis of the posted price. In the checkered history of the international oil industry, there have been three stages in the relationship between producing countries and oil companies regarding the way the former's oil revenues were calculated. In the first stage, the producing countries received only fixed royalties based on the volume of crude produced. The royalties were fixed without regard to the rate of profit which the oil companies eventually took on the crude oil produced. In the second stage, beginning in the 1950s, oil companies were obligated to pay not only the royalties based on the production volume, but also the 50 percent tax under the 50-50 arrangement on the "profits," which were calculated as the difference between the production cost and the posted price.

In the third stage, the producing countries have directly participated in the ownership of the shares of the concessionaire companies set up by the oil companies to exploit oil concessions. The producing country participation was originally known as the 75-25 profit-sharing arrangement, because a producing country and an oil company would each take 50 percent ownership of a concessionaire company. For example, the first 75-25 arrangement was concluded in 1957 between Iran and the Italian oil company ENI. Both the National Iranian Oil Company and ENI took 50 percent interest in a new oil company set up to explore and develop certain Iranian oil fields. Under this

Table 3.1
Development of Posted Prices of Selected Crude Oils of Certain OPEC Countries, 1950-74* (U.S. dollars per barrel)

Date Effective		*Iraq 36° FOB Tripoli*	*Libya 40° FOB Brega*	*Qatar 40° FOB Umm Said*	*Saudi Arabia 34° FOB R. Tannura*	*UAE 39° FOB Jebel Dhana*	*Venezuela 35° Oficina*
Year	*Month*						
1950	Oct. 2	2,410	—	—	—	—	—
1950	Nov. 1	—	—	—	1,750	—	—
1952	July 24	—	—	—	—	—	2,570
1953	Feb. 5	2,290	—	—	1,930	—	—
1953	July 16	2,390	—	2,080	—	—	—
1955	Sept. 1	—	—	—	—	—	2,880
1956	Feb. 9	2,460	—	—	—	—	—
1957	Jan. 15	—	—	—	—	—	3,050
1957	May 28	—	—	2,210	—	—	—
1957	June 7	—	—	—	2,080	—	—
1957	Sept. 10	2,590	—	—	—	—	—
1958	Jan. 31	2,490	—	—	—	—	—
1959	Feb. 6	—	—	—	—	—	2,900
1959	Feb. 13	2,310	—	2,030	1,900	—	—
1959	Apr. 4	—	—	—	—	—	2,800
1960	Aug. 9	2,150	—	—	1,800	—	—
1960	Aug. 16	2,210	—	1,930	—	—	—
1961	Aug. 12	—	2,230	—	—	—	—
1964	Jan. 1	—	—	—	—	1,880	—
1970	Jan. 1	—	—	—	—	—	2,339
1970	Sept. 1	2,410	2,530	—	—	—	—
1971	Jan. 1	—	—	—	—	—	2,344
1971	Feb. 15	—	—	2,280	2,180	2,235	—
1971	Mar. 18	—	—	—	—	—	2,725
1971	Mar. 20	3,211	3,447	—	—	—	—
1971	June 1	—	—	2,387	2,285	2,341	—
1972	Jan. 1	3,136	3,386	—	—	—	3,211
1972	Jan. 20	3,402	3,673	2,590	2,479	2,540	—
1972	Apr. 1	3,373	3,642	—	—	—	—
1972	July 1	3,353	3,620	—	—	—	—
1973	Jan. 1	3,484	3,777	2,705	2,591	2,654	3,362
1973	Mar. 13	—	—	—	—	—	3,768
1973	Apr. 1	3,712	4,024	2,862	2,742	2,808	—
1973	June 1	3,921	4,252	3,025	2,898	2,968	—
1973	July 1	4,070	4,416	3,084	2,955	3,026	—
1973	Aug. 1	4,223	4,582	3,200	3,066	3,140	4,230
1973	Oct. 1	4,243	4,604	3,143	3,011	3,144	4,755
1973	Oct. 16	7,213	—	5,834	5,119	6,045	—
1973	Oct. 19	—	8,925	—	—	—	—
1973	Nov. 1	7,293	9,061	5,899	5,176	6,113	7,110
1973	Dec. 1	7,096	—	5,737	5,036	5,944	7,312
1974	Jan. 1	n.a.	15,768	12,414	11,651	12,636	12,337

*As of January 1 of each year.

Source: Abdul Amir Kubbah, *OPEC: Past and Present* (Vienna, Austria: Petro-Economic Research Center, 1974), pp. 170-71.

arrangement, Iran received 50 percent of the new company's profits plus the usual 50 percent tax on the ENI's half of the profits, resulting in a total take of 75 percent of the new company's profits. The 75-25 arangement is a mixture of the 50-50 arrangement with direct participation in the ownership by a producing country. Of course, if a producing country takes a participation of 70 percent instead of 50 percent, it becomes the 85-15 profit-sharing arrangement.

Under the participation arrangement, there are two different categories of crude oil. "Equity oil" (or "equity crude") is the crude that belongs to the oil companies owning a part of the concessionaire company. Under the 75-25 arrangement, 50 percent of the crude produced is equity oil, while only 30 percent of the crude is equity oil under the 85-15 arrangement. The rest, known as "participation oil" (or "participation crude"), belongs to the producing country. The posted price applies only to equity oil, while the price of participation oil is often expressed as a fixed percentage of the posted price. The amount of oil revenues to a producing country per barrel of crude oil produced depends on the particular mix of these two categories of oil.

The revenue on equity oil to a producing country is derived as follows:

Posted price – (Royalty + Production Cost) = Taxable income

Taxable income × Tax rate = Income tax to a producing country

Royalty + Income tax = Goverment take on equity oil

A producing country can change the government take on equity oil by manipulating either the royalty rate or the income tax rate applicable to equity oil. In contrast, participation oil is either auctioned off by the producing country in a free market or sold back to the oil companies operating the concession. In the latter case, the buy-back price (also called the participation price) is currently set at between 93 to 95 percent of the posted price. Usually the bulk of participation oil is sold back to the oil companies rather than being auctioned off in the free international market for crude oil.

How then does one calculate the average government revenue per barrel of oil consisting of both equity and participation oil? Let us assume that a certain OPEC country owns outright 60 percent of the oil produced (participation oil) and a group of Western oil companies own the remaining 40 percent (equity oil). If the OPEC country sells all its participation oil to the oil companies at a participation price of $10.46 per barrel, and if the government take on equity oil, including the tax and royalties, is $10.12 per barrel,the average government revenue per barrel is $10.32, calculated as follows:

$$\$10.46 \times 60\% + \$10.12 \times 40\% = \$10.32$$

Prices of OPEC oil are based on Saudi Arabian 34° light crude oil, which serves as the reference crude (or the marker crude) for all OPEC oil. Using the Saudi marker crude as an example, let us calculate the average government take per barrel of crude oil produced. Since January 1, 1974, most Persian Gulf members of OPEC have owned 60 percent of the equity shares in the oil companies' concessions. With the 60 percent participation rate, 60 percent of the crude, known as participation oil, belongs to the producing country outright, and the remaining 40 percent, known as equity oil, belongs to Western oil companies. For the first nine months of 1975, the posted price of Saudi marker crude was $11.25 and the buy-back price (or participation price) was set at 93 percent of the posted price. Since the income tax rate of 85 percent applies to the taxable income, while the royalty is 20 percent of the posted price, we can calculate the average Saudi government take per barrel as follows:

Average Government Take Per Barrel of Saudi Marker Crude

(From January 1, 1975 through September 30, 1975)

(1) Posted price	$11.25
(2) Production cost	0.12
Equity oil:	
(3) Royalty payment [20% of (1)]	2.25
(4) Income tax [85% of (1 – 2 – 3)]	7.55
(5) Government take for equity oil [(3) + (4)]	9.80
Participation oil:	
(6) Participation (or buy-back) price [93% of (1)]	10.46

(7) Government take for participation oil [(6) – (2)]	10.34
(8) Average government take [40% × (5) + 60% × (7)]	10.12

At the end of September 1975, the OPEC members agreed to increase by 10 percent the official sales price, or the participation (or buy-back) price, of Saudi Arabian marker crude oil. The increased oil price, which is to remain valid through June 30, 1976, but could be extended until the end of 1976, means a new buy-back price of $11.51 per barrel of Saudi marker crude instead of the previous buy-back (or participation) price of $10.46 per barrel ($10.46 × 110% = $11.51). However, the OPEC Ministerial Conference held in Vienna in late September 1975 did not directly mention any change in the posted price. Since the participation price is 93 percent of the posted price, the new posted price for Saudi marker crude must also have been increased to $12.38 from $11.25 ($11.51 ÷ 93% = $12.38). Therefore, the new average government take may be calculated as follows:

Average Government Take Per Barrel of Saudi Marker Crude

(From October 1, 1975 through June 30, 1976)

(1) Posted price	$12.38
(2) Production cost	0.12
Equity oil:	
(3) Royalty payment [20% of (1)]	2.48
(4) Income tax [85% of (1 – 2 – 3)]	8.31
(5) Government take for equity oil [(3) + (4)]	10.79
Participation oil:	
(6) Buy-back price [93% of (1)]	11.51
(7) Government take for participation oil [(6) – (2)]	11.39
(8) Average government take [40% × (5) + 60% × (7)]	11.15

The new average government take of $11.15 per barrel is an increase of 10.2% over the previous average government take of $10.12 per barrel. But it has become clear that some OPEC member countries do not plan to increase their oil prices by a full 10 percent. For example, both Ecuador and Indonesia have openly announced their intention not to raise oil prices by 10 percent because they have had difficulty in marketing their oil. Some experts estimate that the weighted average increase of all the crude prices may amount to 6.5 to 7 percent instead of 10 percent.[4]

Generally, payments for equity oil are made more quickly than payments for participation oil, because OPEC countries usually allow the oil companies a longer period of credit for participation oil payments than for equity oil payments. During the second half of 1974, equity oil payments represented about 43 percent of the total oil revenues to OPEC countries.[5] While the details on arrangements for participation oil payments are mostly unknown, the following schedule for equity oil payments suggests that most participation oil payments may be made with a lag of up to 90 days after the end of a quarter. This may be the reason why it is only within the last few months of 1974 that the OPEC countries accumulated an increased amount of oil dollars, even though the posted prices had been raised steeply by January 1, 1974.

Due Date of Roylty and Tax Payments on Equity Oil

Kuwait	20 days after end of quarter
Nigeria	20 days after end of quarter
Abu Dhabi	30 days after end of quarter
Libya	60 days after end of quarter
Iran	15 days after end of quarter
Saudi Arabia	Royalty is due 15 days after end of previous month. Tax is due 45 days after end of quarter.

Source: "Economics and Portfolio Strategy," Paine, Webber Jackson & Curtis, Inc., Nov. 22, 1974.

This payments delay stems from the oil companies' practice of taking oil on consignment. While some OPEC countries have on occasion tried to accelerate the payments, they appear to have met with only modest success, partly because there are certain OPEC countries which sometimes extend the payments period as an indirect way to increase their crude sale by offering better payments terms. Even though the oil price was more than quadrupled in 1973, it was for this reason that the actual payments were slow in coming. For example, it was not until well into the first quarter of 1974 that the OPEC countries began to receive the increased oil revenues which resulted from the autumn

1973 price raises. Receipts from the end-1973 raise in oil prices started to come in only in the second quarter of 1974.[6]

Recently, there have been some movements to change the whole system of determining the government take. On December 13, 1974, at the end of the 42nd OPEC Ministerial Conference held in Vienna, it was announced that the OPEC members would soon abolish the posted price; there would be a fixed government take instead of a combination of royalty and tax on equity oil. As for the participation oil price, the OPEC spokesman did not clarify the situation, but it was assumed that the participation price would also be fixed independently of the posted price, thus abolishing the need for a posted price entirely. In addition, it was decided at the OPEC Conference to reduce the participation oil price slightly, while the equity oil cost was increased quite a lot due to an increase in the government take on equity oil. The result is that the bonus of equity oil over participation oil is reduced for an oil company, as shown in the following table (for Saudi Arabian light crude, $ per barrel):

	Old	New
(1) Participation oil price (1)	$10.83	$10.46
(2) Government take on equity oil (2)	9.74	10.12
(3) Production cost (3)	0.10	0.12
(4) Equity oil cost (2) + (3)	9.84	10.24
(5) Equity oil bonus (1) – (4)	$0.99	$0.22

OPEC's reason for reducing the equity oil bonus might be to pressure Western oil companies to accede to the demands of certain OPEC members for a complete takeover of concessionaire companies. The incentive for oil companies to have equity ownership will decline as the bonus of equity oil over participation oil is reduced. Theoretically, if the equity oil bonus declines to zero, Western oil companies will become indifferent to the distinction between equity oil and participation oil, because there will be no price advantage for oil companies in getting equity oil instead of participation oil.

PROSPECTS FOR OIL PRICE MOVEMENTS

The future development of oil prices will be one of the most decisive factors in establishing the long-term government take of the OPEC countries. The best way to consider prospects for OPEC oil prices is to study the supply and demand curves for oil. Before 1973, the supply curve of crude oil was a highly elastic one, while the demand curve was (and still is) highly inelastic, at least in the short run. Throughout the 1960s, the crude price of about $2 a barrel was much like the price a competitive industry would have charged, by producing and selling so long as the market price is above marginal costs. The direct production costs of crude oil in the Middle East were as low as 10 cents per barrel in the 1960s, although they went up to 30 cents a barrel by the early 1970s. Even at $2 a barrel, however, oil was *at least* three times cheaper than its nearest substitute, coal.

So there existed, throughout the 1960s, a substantial "rent" that could have been obtained by pricing oil up to a level just below the price of substitutes such as coal. This rent could have been collected by the producing countries as a monopoly profit, but despite the formation of OPEC in 1960, they did not succeed in doing so until 1973. Earlier, we explained the reasons for this delay. In any case, the inability of producing countries to exploit the monopoly profit had the consequence that the governments of consuming countries collected part of the rent in taxes from oil consumers, hence redistributing the income internally. For example, by September 1973, the governments of Western Europe were taxing about $7.50 of the average $14.50 a barrel price of refined oil products in Western Europe. Figure 3.1 shows the supply and demand curves for the crude oil market until 1973, when OPEC members quadrupled oil prices. In the figure, it is assumed that the supply curve (SS_1) before 1973 was highly price-elastic; the crude-oil market was very competitive due to aggressive marketing by non-major (or independent) oil companies that were largely blocked from selling their crude in the U.S. market because of the mandatory oil-imports quota system. It is also assumed that the short-run demand curve (D_SD_S) is highly price-inelastic, while the long-run demand curve (D_LD_L) is assumed to be less price-inelastic than the short-run demand curve, because other sources of energy, such as coal, can be increasingly used as the price of oil goes up.

Figure 3.1
Crude Oil Market before 1973

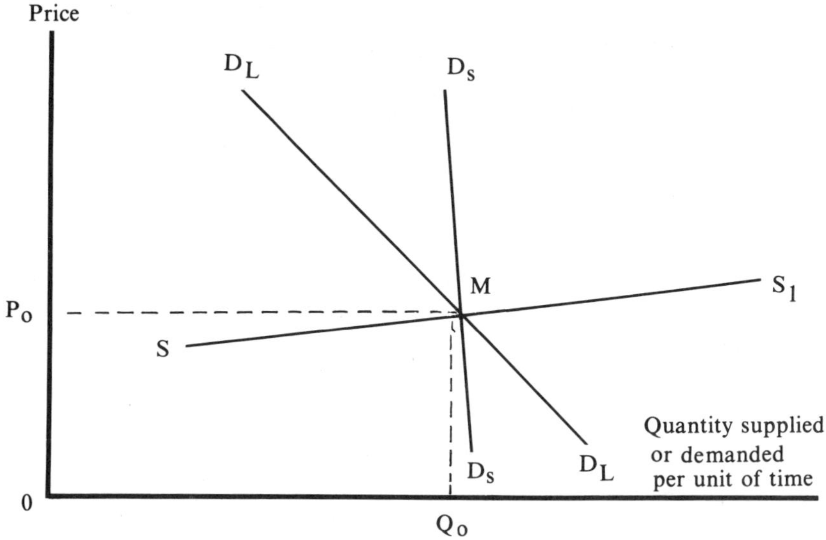

D_LD_L: long-run demand curve
D_sD_s: short-run demand curve
SS_1: crude-oil supply curve before 1973

The short-run price inelasticity of the demand curve was fully exploited by the quadrupling of oil prices by OPEC members in 1973. Even though the price of oil climbed by more than 400 percent in 1973 and 1974, the demand for oil declined by only 10 percent or so.

Figure 3.2 (p. 38) shows the new oil market condition after the 1973 oil-price increases, as indicated by the shift of the supply curve from SS_1 to SS_2. The new supply curve (SS_2) is an oligopolistic supply curve solely determined by OPEC members themselves. Just like any oligopolists, however, OPEC members will attempt to set the oil price at the level where they can maximize their revenues, and, to that extent, their oligopolistic supply curve (SS_2) is also market-determined. Admittedly, the shape of SS_2 in Figure 3.2 is conjectural and not based on any econometric test of OPEC supply behavior. It is presented here

Figure 3.2
Oil Market after the 1973 Price Increases

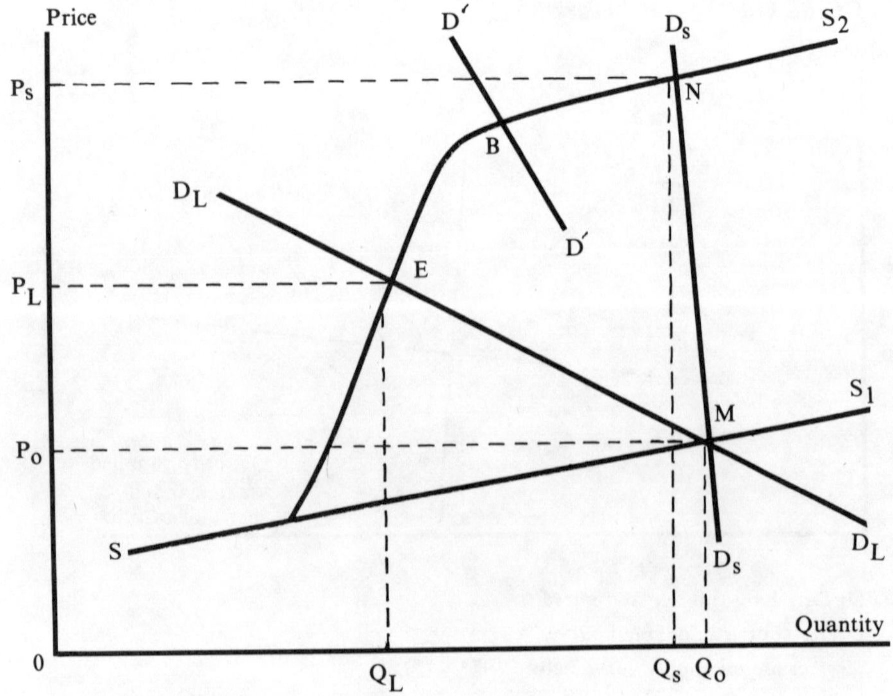

SS_2: OPEC "supply curve" since 1973, oligopolistically determined, thus dependent upon the demand curve also.
B: OPEC cartel breaking point

simply as an illustration of a possible shape of the post-1973 supply curve. Over the short term, the oil price (P_S) is determined by the intersection of SS_2 and D_SD_S at the point N. Even though the oil price has quadrupled ($OP_S = 4 \times OP_O$), the quantity has declined by only 10 percent ($OQ_S = 0.9 \times OQ_O$). In the long run, however, the oil price will be determined by the intersection of SS_2 and D_LD_L, the long-run demand curve. As the demand curve gradually shifts from D_SD_S toward the direction of D_LD_L, OPEC member countries may be subject to an increasing internal strain. The demand curve shift may be accompanied by a sharp decline in output, as the quantity demanded moves from OQs toward OQ_L.

In 1974, OPEC members were able to prevent any serious price erosion by voluntarily cutting back oil production in the face of a declining demand caused by both high oil prices and world economic recession, as seen in the following table (Table 3.2).

Table 3.2
Production Cutbacks of OPEC Members in 1974
(million barrels per day)

	Capacity	Production		
Country	1974	Sept. 1973	Dec. 1974	% Change
Saudi Arabia	9.49	8.57	8.05	- 6%
Iran	6.00	5.83	5.94	+ 2
Iraq	2.39	2.11	2.18	+ 3
Kuwait	3.70	3.53	2.32	-34
Qatar	0.64	0.60	0.52	-13
Abu Dhabi	1.90	1.40	1.21	-14
Venezuela	3.05	3.39	2.70	-20
Libya	2.30	2.29	1.00	-56
Algeria	1.10	1.10	0.90	-18
Nigeria	2.40	2.14	2.10	- 2
Indonesia	1.50	1.42	1.10	-23
Total	34.47	32.38	28.02	-13%

Sources: *Petroleum Intelligency Weekly*, and *The Financial Times*, February 24, 1975.

In 1975, there was a further cutback in oil production. World oil consumption, which fell in 1974 at rates varying from 3 percent in Japan and 4 percent in the United States to up to 10 percent in the European nations, declined still further by several percentage points in the first half of 1975, as the economic recession deepened. In order to keep crude prices up in the face of declining demand, OPEC members cut their combined oil production to

26.5 million barrels per day in June 1975, a volume which was substantially lower than the pre-embargo production of 32 million barrels per day in September 1973, and barely 70 percent of the combined production capacity of 38 million barrels per day as of June 1975.

Among the OPEC members, the countries which have a low absorptive capacity with small populations and high per capita incomes (such as Kuwait and Saudi Arabia) may be more willing (and able) to cut down their output to keep up prices. On the other hand, the countries which have large populations and low per capita incomes (such as Indonesia, Nigeria and Venezuela) will be less willing, and politically less able, to cut back oil production drastically. They may be increasingly tempted to "cheat" or undercut other OPEC members to maintain their oil revenues. We may already be witnessing such developments to some extent.[7]

Therefore, we can asume that there is a "kink" or "cartel-breaking point" (B) in the supply curve in Figure 3.2. The exact location of B on the supply curve will depend on the degree of solidarity of OPEC. As the shift leftward of the demand curve reaches D'D' to intersect the supply curve SS_2 at B, the OPEC cartel may break up. Thereafter, the supply curve may assume a steeper slope, and the long-term equilibrium price may be achieved at E, where the long-term demand curve (D_LD_L) intersects the supply curve (SS_2). The leftward shift of the demand curve is due to two factors. First, the high oil price encourages more effective conservation measures by consumers, Second, more oil substitutes such as coal, solar energy, nuclear energy, and geo-thermal power will be offered on the market, since oil has become a rather expensive energy source at the present prices.[8]

OPEC members froze oil prices for nine months from January 1 through September 30, 1975, when they raised them again by 10 percent. They also argued that future oil prices would have to be increased further *pari passu* with the world inflation rate. In this connection, many OPEC countries have spoken in favor of "indexation" of future oil prices to the prices of major imports of

OPEC countries. In contrast, oil-consuming countries, notably the United States, argue that the current oil prices are already too high, that a stable equilibrium based on supply and demand requires a substantial roll-back of present oil prices. But outright price roll-backs are highly unlikely and oil prices in the immediate future will be determined largely by the relative strengths of the bargaining positions of oil-producing and oil-consuming countries. In the long run, though, the equilibrium market forces will determine the oil prices in a way shown by Figure 3.2. How long it will take to arrrive at point E will depend on the shape of the OPEC "supply curve," which in turn depends on the degree of solidarity among OPEC members. For example, if negotiations within OPEC on the sharing of production cut-backs should be successful, oil prices could be held up for a long time; in this case, the supply curve could remain almost horizontal at the OPEC-administered price. In order for such negotiations to be successful, the OPEC members with a low absorptive capacity must be willing to cut down their oil production relatively more than the members with high absorptive capacity. Up to now, however, this has not necessarily been the case, as can be seen in Table 3.3.

Table 3.3
Production Cutbacks of OPEC members in the First Half of 1975 (million barrels per day)

	Production			
Country	June 1975	% Change from 1974	Prod. Capacity June 1975	% Production of Capacity
Saudi Arabia	6.8	-23%	10.3	66%
Iran	5.2	-16	6.5	80
Kuwait	2.2	-23	3.8	58
Iraq	2.3	+29	2.6	88
U.A.E.	1.9	+ 3	2.2	86
Qatar	0.3	-36	0.7	43
Venezuela	2.4	-18	3.3	73
Nigeria	1.6	-32	2.5	64
Indonesia	1.1	-25	1.6	69
Libya	1.5	-11	3.0	50
Algeria	0.8	-33	1.1	73
Gabon	0.2	+12	0.2	100
Ecuador	0.2	-31	0.3	67
Total	26.5	-17%	38.1	70%

Source: Based on data from *Petroleum Intelligence Weekly* and *The Financial Times*, August 22, 1975.

The solidarity of OPEC depends on a number of factors. According to Waverman, a cartel requires four esential factors to be successful and durable. First, there must be a low elasticity of demand for the industry's product. Second, there must be a high elasticity of demand for the product of any one producer, because the nature of the product is homogeneous and perfectly non-differentiated. Since any producer can easily outsell all other producers by slightly lowering the price of his own product, all the producers have a strong common interest in maintaining group solidarity in order not to fall into cutthroat competition. Third, the producers should have similar costs. Fourth, there should be a small number of firms (or countries) in the cartel.[9] According to the above criteria, the solidarity of OPEC rests on a firm foundation, and the cartel may be able to maintain its oligopolistic pricing system for some time to come. But its main weakness lies in the high degree of heterogeneity in the political and economic structures of the OPEC members. This weakness was especially evident during the recent OPEC Ministerial Conference, held in September 1975, at which extreme difficulty was encountered in reaching the 10 percent price increase agreement.[10]

NOTES

[1]E.T. Penrose, *The Large International Firm in Developing Countries* (London: George Allen and Unwin, 1968), p.43.

[2]Taki Rifai, *The Pricing of Crude Oil* (New York: Praeger Publishers, 1974), p. 33.

[3]See, for example, W. L. Nelson, "Sulfur Contents of U.S. Crude Oils," *Oil and Gas Journal*, August 17, 1970; H. J. Frank, *Crude Oil Prices in the Middle East* (New York: Praeger Publishers, 1966); W. A. Leeman, *The Price of Middle East Oil* (Ithaca, N. Y.: Cornell University Press, 1962.

[4]"Oil-Price Rise Will Not Equal 10% OPEC Set," *The Wall Street Journal*, September 30, 1975; and "Yamani: Oil Prices Rose by Only 6.5%," *The Journal of Commerce*, October 8, 1975.

[5]Joseph P. Carroll, Vice President of the Chase Manhattan Bank, "Institutional Concerns in Managing Petro-dollar Investments," (Speech at the World Trade Institute Symposium on Investment Alternatives for the New Oil Money, New York, October 23, 1974.)

[6]*International Review* (Lionel D. Edie & Company, New York, February, 1975), p. 9.

[7]See, for example, "Oil-Price Shaving Is Spreading in OPEC; Broad 'Cheating' by Members Seen Next," *The Wall Street Journal*, July 18, 1975.

[8]For elaborations of these points, see Robert Aliber, "Impending Breakdown of OPEC Cartel," *The Wall Street Journal*, March 20, 1975.

[9]Leonard Waverman, "Oil and the Distribution of International Power," *International Journal*, August 1974, p. 625.

[10]For an evaluation of the bitter internal arguments among the OPEC members and their future price implications, see Adrian Hamilton, "End of Old Assumptions on Oil Prices: Not the Oil Price Increase Itself, But the Way the Decision Was Reached, May Be Most Important for Consumers," *The Financial Times*, September 29, 1975.

Chapter 4

Oil-Money Surpluses and the Balance of Payments Adjustment Problem

ESTIMATES OF OIL-MONEY SURPLUSES

In the wake of the quadrupling of oil prices in 1973, the initial reaction of both economists and government officials was directed to the prospect the OPEC countries would accumulate enormous sums of oil-money surpluses, with serious implications for the world's balance of payments. Early estimates of oil-money surpluses to be accumulated by OPEC members were as high as $1 trillion or more by the year 1985. For example, one of the earlier World Bank estimates put the cumulative buildup of funds by the OPEC countries at $650 billion by the end of 1980 and at $1.2 trillion by the end of 1985. The concentration of so much money in the hands of a relatively small number of countries within ten years or so could have serious effects on the existing world economic order in terms of real income, balances of payments, and the international monetary system.

Since early 1974, a number of other economists have estimated the expected OPEC surpluses, with widely varying results. These variations are due to different assumptions regarding the factors that will determine the level of OPEC surpluses. In general, there are five crucial factors involved here: the volume of OPEC oil exports; the average OPEC government take per barrel of oil produced; the volume and price of OPEC imports from non-OPEC countries, as well as of OPEC non-oil exports; the magnitude of grant aids by the OPEC countries to non-OPEC countries; and the average rate of return on OPEC external investments.

Since late 1973, we have witnessed two main stages in the evolution of oil-money projections. During most of 1974, the prevailing view was that the OPEC countries as a group would accumulate vast sums of oil-money surpluses, which would seriously disrupt the world economic order unless bold remedial actions were taken. Since late 1974, however, several developments have inspired more optimistic oil-money projections. First, the rate of growth of OPEC oil exports has slowed down and then even declined slightly, due to a combination of factors such as a mild 1974/75 winter, the economic recession in the Western countries, and, to a certain extent, the depressive effect of oil-price rises on the amount of oil consumption. Second, the OPEC countries have stepped up their imports from non-OPEC countries much more sharply than had generally been foreseen. For example, the total value of imports by the OPEC members rose about 80 percent in 1974 over that in 1973. Price increases on an average accounted for somewhere around 25 percent of this increase. Third, the OPEC countries have learned relatively quickly how to invest a growing portion of their surplus revenues in long-term development projects both within their own countries and abroad, thus reducing the potential amount of financial assets to be accumulated by them. Finally, as will be discussed in detail in Chapter 9, many OPEC countries have increased financial aid to certain developing countries, thereby reducing their reserve accumulation.

The balance-of-payments data for 1974 show striking contrasts between the OPEC members. Three countries, Saudi Arabia, Iran and Kuwait, account for almost two-thirds of the OPEC surplus. Of the remaining members only Nigeria, Venezuela, and the United Arab Emirates (UAE) had significant surpluses. This financial disparity was primarily the result of different levels of oil production. Although imports increased markedly in all OPEC countries, there were significant differences among them. At the extremes, the imports of Iraq and Iran increased nearly 200 percent and 125 percent respectively, while the imports of Libya and Nigeria rose less than 40 percent in nominal terms. The following two tables, Table 4.1 and Table 4.2, show the financial surpluses of twelve OPEC members, excluding Gabon.

Table 4.1
OPEC Investible Surplus in 1973 (in millions of U.S dollars)

	Oil Exports (Gov't Take)	Non-oil Exports	Imports f.o.b.	Investible Surplus
Algeria	1,000	360	2,060	- 870
Ecuador	100	310	460	- 40
Indonesia	1,200	1,610	2,410	- 350
Iran	4,500	590	3,600	1,090
Iraq	1,700	110	1,160	470
Kuwait	1,900	230	920	1,520
Libya	2,300	-	2,200	- 600
Nigeria	2,400	620	1,780	260
Qatar	400	10	180	140
Saudi Arabia	5,500	20	1,800	3,120
U.A.E.	1,200	40	860	290
Venezuela	3,000	375	2,820	- 135
Total	25,200	4,275	20,250	4,895

Source: U.S. Treasury Department, "The Absorptive Capacity of the OPEC Countries," Washington, D.C., September 5, 1975.

Since late 1974 there have appeared more optimistic assumptions on the factors determining the level of cumulative surpluses, leading to new projections which are much lower than the earlier ones. For example, a study published by the Morgan Guaranty Trust Company of New York in early 1975 projects that OPEC's accumulations of financial assets would peak at $248 billion by the end of 1978. Their projection is based on the following assumptions: little change for the remainder of the decade in demand for OPEC oil by the rest of the world; a 5 percent per annum average increase in the OPEC governments' take per barrel; an average 15 percent per annum increase in OPEC non-oil exports; a 20 percent average annual increase in the volume of OPEC imports; a 12 percent increase in the price of OPEC

Table 4.2
OPEC Investible Surplus in 1974 (in millions of U.S. dollars)

	Oil Exports (Gov't Take)	Non-oil Exports	Imports f.o.b.	Investible Surplus
Algeria	3,700	355	3,710	405
Ecuador	500	450	790	110
Indonesia	3,400	2,200	3,890	230
Iran	18,700	800	8,000	10,680
Iraq	5,700	150	3,460	1,970
Kuwait	8,000	390	1,480	7,355
Libya	6,200	40	3,000	2,540
Nigeria	7,600	850	2,490	5,220
Qatar	1,600	10	270	1,290
Saudi Arabia	24,600	25	3,530	20,800
U.A.E.	6,000	20	1,600	4,360
Venezuela	8,900	375	4,660	3,995
Total	94,900	5,665	36,880	58,935

Source: U.S. Treasury Department, "The Absorptive Capacity of the OPEC Countries," Washington, D.C., September 5, 1975.

imports in 1975, and an average 7 percent increase per year thereafter; and an 8 percent per annum financial return on OPEC external investments.[1] The Morgan study also projects that the OPEC countries as a group would be running trade *deficits* of $29 billion in 1979 and $69 billion in 1980, thereby reducing the outstanding OPEC financial surpluses to under $180 billion by the end of 1980. (The Morgan projections, like other projections to be discussed in this chapter, are in current dollars, unless stated otherwise.)

A similarly optimistic projection was made by the Irving Trust Company of New York in March 1975. The bank develops two different scenarios based on two different projections of future oil prices: one in which nominal oil prices rise by 20 percent in 1975, 10 percent in 1976, and 7 percent per annum thereafter, matching inflation; and the other in which oil prices start to fall in 1977. In the first scenario, the cumulative financial surpluses of OPEC countries as a group would peak at $265 billion at end-1979 and then would decline to $248 billion at the end of 1980. In the second scenario, the cumulative surpluses are projected to peak at

Table 4.3
Morgan Guaranty Projections of OPEC Surpluses
(in billions of U.S. dollars)

	1974	1975	1976	1977	1978	1979	1980
Oil revenues	105	110	119	125	128	135	143
Trade balance	62	52	44	27	1	-29	-69
Current-account balance	65	60	57	43	20	-10	-53
Cumulative external financial assets	80	137	191	231	248	235	179

Source: *World Financial Markets,* Morgan Guaranty Trust company of New York, January 21, 1975.

Table 4.4
Irving Trust Projections of OPEC Surpluses
(in billions of U.S. dollars)

Case A: Oil Prices Rise	1974	1975	1976	1977	1978	1979	1980
Oil revenues	108	119	120	116	124	132	141
Merchandise trade balance	83	77	61	42	31	14	- 8
Current-account balance	65	64	51	34	24	7	-17
Cumulative financial surplus	85	149	200	234	258	265	248
Case B: Oil Prices Break in 1977							
Oil revenues	108	119	120	90	81	72	63
Merchandise trade balance	83	77	61	16	-12	-46	-86
Current-account balance	65	64	51	8	-21	-59	-106
Cumulative financial surplus	85	149	200	208	187	128	22

Source: *The Economic View from One Wall Street,* Irving Trust Company, March 20, 1975.

$208 billion in 1977, then decline to only $22 billion by the end of 1980.

A third optimistic projection was made by the First National City Bank of New York in mid-1975 (Table 4.5). The bank's "central scenario" has OPEC running a current-account deficit of $7 billion in 1980 and its cumulative surpluses peaking at $196 billion at the end of 1979. In contrast to these relatively optimistic projections, there are some recent studies which still project substantial oil-money surpluses. For example, Walter Levy, the eminent oil economist, takes a strong exception to the "premature" optimism reflected in the projections mentioned above. First of all, he believes that oil prices are not likely to break easily. Consequently, the cumulative financial surpluses of the OPEC countries would reach about $450 billion by 1980. This figure is similar to the one published by the World Bank in April 1975.

Table 4.5
FNCB's "Central Scenario" Projection of OPEC Surpluses (in billions of U.S. dollars)

	1974	1975	1976	1977	1978	1979	1980
Oil revenues	126	107	116	114	113	108	103
Trade balance	70	42	42	32	23	11	- 4
Current-account balance	66	36	37	30	19	8	- 7
Cumulative financial surplus	66	102	139	169	188	196	189

Source: *Money International,* First National City Bank, May 30, 1975.

Table 4.6
OPEC Surpluses Projected by W.J. Levy (in billions of U.S. dollars)

	1974	1975	1976	1977	1978	1979	1980
Oil revenues	95	95	123	138	152	162	168
Trade balance	57	43	62	62	58	45	23
Cumulative surpluses	75	122	191	264	337	402	449

Source: *Future OPEC Accumulation of Oil Money: A New Look at a Critical Problem,* W.J. Levy Consultants Corp., New York, June 1975.

So far, we have reviewed several well-known studies projecting the amount of financial surpluses to be accumulated by the OPEC countries. Regardless of whether some studies project larger cumulative surpluses than others, it is quite clear that the world economy is faced with the necessary redistribution of a vast sum of financial resources amounting to hundreds of billions of dollars. In this circumstance, the most urgent economic issue is how to "finance" the huge oil deficits. This is a balance-of-payments problem for most oil-importing countries, which are forced to suffer a sudden and enormous increase in their trade deficits. Oil is such a strategic and essential commodity for the modern economy that, during the coming several years at least, most oil-consuming countries have no other alternative than to continue to import oil, no matter what its cost for the moment. For most oil-importing countries, especially oil-importing LDCs, how to finance increased oil-import bills is a critical problem. This question, however, will be discussed in detail in Chapter 8, which deals with the oil-money recycling problem.

CONVENTIONAL ADJUSTMENT THEORIES AND OIL-INDUCED IMBALANCE

Along with the oil-financing problem, most oil-importing countries also face the basic task of balance-of-payments adjustment in view of huge oil-related deficits. Due to the quadrupling of oil prices, we are witnessing the severest disturbance in world trade balances since the Second World War. Whether the combined OPEC surplus per year is $40 billion or $60 billion, and whether the cumulative OPEC surpluses by 1980 will amount to $200 billion or $400 billion, the fact remains that the oil deficits pose one of the most difficult balance-of-payments adjustment problems in the international monetary system. Both the size and the suddenness of the oil-price increases have already generated severe disruption in international trade relationships. The huge trade deficits of many oil-importing countries, and the even greater trade surpluses of several OPEC countries, present especially thorny policy issues fundamentally different from those normally associated with the process of international adjustment.

Conventional theories of balance-of-payments adjustment envisage exchange-rate changes and/or incomes policies to deal with persistent trade imbalances. Changes in exchange rates will affect relative prices of imports and domestic import-competing goods on the one hand, and of exports and export-competing goods in the rest of the world, on the other. How much the foreign currency proceeds of a country's exports will change depends upon the elasticity of foreign demand for the country's exports and the elasticity of domestic supply of export goods; similarly, on the import side, it depends upon the elasticity of domestic demand for imports and the elasticity of foreign supply of imports.[2] Therefore, the extent to which a change in an exchange rate contributes to adjustment toward a country's overall trade balance depends on the four price elasticities for both domestic and foreign demand and supply of imports and exports. A change in exchange rates can contribute to trade adjustment only when it satisfies the Marshall-Lerner condition that, under the assumption of initial trade balance and infinite supply elasticities, the sum of elasticities of demand for imports in two countries exceeds one.

Apart from the controversy over elasticity pessimism or optimism, which centers on the probability of the Marshall-Lerner condition being fulfilled in reality, the elasticity or relative-price approach is patently ill-suited as an adjustment mechanism to the oil-deficit problem for two reasons. First, oil prices are expressed in U.S. dollars, (and, to a limited extent, in British pounds) so that upvaluation of OPEC's currencies will not necessarily affect the amount of oil deficits. Second, oil is an essential commodity for a modern economy, and it is highly price-inelastic within a time span of at least several years. Even if oil prices were further increased by a devaluation of oil-importing countries' currencies, there would be no appreciable decline in the demand for oil; any increase in oil prices would only lead to further harm to the economy of an oil-importing country in terms of inflation, "excise-tax" effect, and worsening terms of trade.

Apart from the relative-price or elasticity approach, a new theory of adjustment, the aggregate-spending or absorption theory, has developed since the early 1950s. The elasticity approach assumed, explicitly or implicitly, the existence of unemployed resources that could absorb the multiplier effects of devaluation-induced changes in the demand for domestic output. However, this theory proved extremely unsatisfactory in the post-war situation of general inflationary pressure. Unlike the elasticity approach, the essence of the absorption approach, formalized by Sidney Alexander,[3] is to view the balance of payments as a relation between the aggregate receipts and expenditures of the economy, rather than as a relation between the country's credits and debits on international account. Alexander criticizes the elasticity approach as an extension, to imports and exports as a whole, of the Marshallian partial equilibrium analysis of the price and production of a single commodity. While the Marshallian supply and demand curves are very useful tools for analyzing the factors that determine price and output for a single good, he questions their value when applied to imports and exports as a whole. Consequently, Alexander proposes to view the foreign trade balance, B, as the difference between a country's total production of goods and services, Y, and its total "absorption" of goods and services, A:

$$B = Y - A \quad \ldots\ldots\ldots\ldots \quad (1)$$

The above equation is based on the fact that a country's trade balance is equal to the difference between the total goods and services produced in that country and the total goods and services taken off the market domestically. "Absorption" then refers to the taking of goods and services off the market, and equals the sum of consumption and investment (including inventory changes). By denoting changes in the variables in Equation (1) by the corresponding small letters, Alexander expresses his "fundamental identity" as follows:

$$b = y - a \quad \ldots\ldots\ldots\ldots \quad (2)$$

Absorption of goods and services depends on real income, which itself is equal to the output of goods and services, and on the price level, which is affected by the change in exchange rates. Therefore, we may have a functional equation

$$a = cy - d \; .... \quad (3)$$

where c is the propensity to absorb and d the direct effect of the exchange-rate change on absorption. A combination of the functional relationship (3) with the fundamental identity (2) yields:

$$b = (1-c)y + d \; ...... \quad (4)$$

Equation (4) makes it clear that the effect of an exchange-rate change on trade balance is determined by that change's effect on c, y, and d. Thus, the absorption approach "poses the problems of balance-of-payments in a way which highlights their macro-economic policy implications, and it allows for conditions of full employment and inflation."[4] The absorption approach shows us that, if an improvement of the foreign trade balance is desired, policies should be adopted so as to reduce absorption relative to production.

Like the elasticity approach, however, the absorption approach cannot be usefully applied to the oil-deficit problem. First of all, those OPEC member countries that experience huge trade surpluses are unable to increase their absorptive capacity due to their unique economic structures characterized by small populations, scarcity of both skilled manpower and technical-managerial expertise, etc. No matter how far expansionary economic policies were to be pursued by these oil-exporting countries to increase their absorption, there would be little prospect that it would increase quickly to a level sufficient to contain most oil-connected surpluses. On the other hand, oil-importing countries may try individually to lower their absorption relative to their domestic production through such measures as deflationary policies, devaluations, trade barriers etc., but this may only lead to a collective economic recession in all oil-importing countries, as well as to trade wars among themselves, rather than lessen to any significant extent their overall trade deficits as a group vis-a-vis oil-exporting countries.

A DYNAMIC ADJUSTMENT MODEL

It is clear, therefore, that conventional adjustment measures such as devaluation and/or deflationary policies are not applicable to the current oil-deficits problem.[5] Consequently, we have to develop a new adjustment approach to deal with the critical balance-of-payments problem posed by the quadrupling of oil prices. Conventional adjustment mechanisms such as exchange-rate changes and incomes policies are patently inadequate, because oil is highly price-inelastic and is paid for in the currencies of some of the oil-importing countries rather than in the currencies of the oil-exporting countries, and because both institutional and economic structural limits in the absorptive capacity of many oil-exporting countries render them unable to spend all their foreign-exchange earnings on current imports within a reasonably short period. Even if they could spend them, there would be a limit to how much a surplus oil-exporting country should be asked to suffer inflation or, on the other hand, a deficit oil-importing country asked to suffer deflation, just to achieve external equilibrium.

Consequently, we come to realize that the crucial element needed to solve the adjustment problem caused by the current oil crisis is time, that is, a reasonable time period over which payments adjustment may be spread out in order to achieve gradual trade equilibrium between oil-exporting and oil-importing countries without serious economic disruptions. The ultimate solution to the current oil-induced adjustment problem would consist in raising the overall absorptive capacity of the oil-exporting countries to the level where their absorption matches, or even exceeds, their domestic production, including huge oil-related export surpluses. This, of course, will take time. Consequently, an adjustment model which would be relevant to the current oil-induced adjustment crisis should take the time element into consideration. But both conventional adjustment theories, the elasticity and absorption approaches, are devoid of a time element; they are essentially static models. Nevertheless, payments adjustment is an economic-growth phenomenon, involving a critical trade-off between economic growth and trade adjustment.[6] What is now needed, therefore, is a dynamic adjustment model which explicitly recognizes and incorporates the time element in the context of economic growth and payments adjustment.

While most components of the pure theory of international trade are quite old, the theory of economic growth and international trade has been developed more recently, within the past two decades. As the starting point for the modern theory of growth and trade, one usually turns to a paper by J. R. Hicks published in 1953.[7] The topic Hicks dealt with in his paper was the long-run development of Great Britain's terms of trade and balance of trade. In order to explain a secular change in the terms of trade, Hicks classifies three different types of economic growth, depending upon whether productivity grows uniformly in all industries, in the export sector, or in the import-competing sector. These different types of growth are respectively called *neutral* growth, *export-biased* growth and *import-biased* growth.

According to Hicks, export-biased growth is an "immiserizing growth" in which the fruit of productivity growth is lost to a country owing to its worsening terms of trade. The case of immiserizing growth was discovered anew in the late 1950s by H. G. Johnson and J. Bhagwati.[8] In the early 1960s, Bo Soedersten also developed a model to study the effects of economic growth on international trade in a more thorough manner than did Hicks and others.[9] However, the main concern of these early "growth and trade" economists was to analyze the effect of economic growth on the terms of trade, assuming that external balance had already been achieved. In other words, their theory of growth and trade assumed an automatic external adjustment, and any problem of external imbalance arising from economic growth was assumed away. Rather, they were interested in finding out how a country, or certain sectors within a country, which experienced economic growth, would fare economically relative to other countries.

In contrast, our intention in this study is to find out how economic growth affects external balance itself. The problem of external balance is not assumed away. We are interested in evaluating the effect of economic growth on the process of adjustment toward external balance. Accordingly, unlike the traditional theory of growth and trade, we do not assume automatic and instantaneous adjustment. On the contrary, external disequilibrium is considered as an integral part of economic growth.

Another strain in the theory of growth and trade has been concerned with the interrelationship between economic growth and trade balance. Its main interest lies in finding out whether economic growth leads to a negative or positive trade balance.[10] Both a simplified Harrod-Domar model of growth in an open economy and a simplified neoclassical growth model with a foreign sector have been used to show that economic growth generally results in trade deficit, or, conversely, that trade deficit stimulates economic growth.[11] This conclusion runs counter to empirical observations, however, for in the 1950s and 1960s, some of the most rapidly growing industrialized countries, such as Japan and West Germany, experienced substantial trade surpluses. In contrast, slowly-growing countries such as Great Britain and the United States have had much less favorable trade experiences.

These empirical observations raise the opposite argument that a positive trade balance does serve as an important engine of growth. As opposed to the trade pessimists' view of growth as leading to income rises and spillovers that expand imports more rapidly than exports, the optimists stress the innovational aspects of growth, and the tendency of both new products and cost-reductions for existing lines to foster exports and enhance import-competing production.

However, the empirical validity of the trade pessimists', as well as the trade optimists', view of growth depends on the relative degree of the income elasticities of imports and exports at the particular stages of a country's economic development. For example, countries which have just started to develop their economies by opening up mines, constructing factories, establishing new farms, etc., generally have a high income elasticity of demand for imports, owing to a great need to bring in capital goods and technology from abroad. On the other hand, their exports generally have a low income elasticity of demand from foreigners, because they tend to consist of primary commodities and not of industrial goods of fine quality that are eagerly demanded by generally affluent foreign consumers. For these developing countries, economic growth usually produces a negative influence on trade balance.

If a country has reached a certain threshold level of economic maturity, its income elasticity of demand for imports may start to fall as more import-competing goods are produced domestically, and as the need lessens for importing all sorts of capital goods and technology. At the same time, the income elasticity of its exports may rise, as more of its domestically produced industrial goods gradually become acceptable to foreign customers. As a result, the country is likely to have a positive correlation between economic growth and positive trade balance. After many years, however, the country may gradually lose its earlier initiative and vitality in foreign trade to new, emerging countries, as its workers lessen its competitive advantage by demanding higher and higher wages. At this stage, the country's growth may again cause trade deficit rather than trade surplus.

As can be ascertained in the above discussions, the main interest of "growth and trade" theorists in this case is in studying the effect of economic growth on trade balance or, conversely, the effect of trade balance on growth. In fact, the models developed to explain the above phenomena, such as the simplified Harrod-Domar model and simplified neoclassical models, are essentially growth models that have explicitly incorporated the foreign sector. The primary focus of these models is on the growth phenomenon itself. In contrast, our concern here is to develop an adjustment model in order to explain the process of external adjustment in a dynamic, growth-oriented context. Therefore, even though we make use of the concept of interrelationship between economic growth and foreign trade, our frame of reference is fundamentally different from that of the conventional theory of growth and trade. We are interested not in the phenomenon of economic growth itself, but in the process of trade adjustment in the context of economic growth. We have developed a model for this process, described in detail at the end of this chapter in the section called Technical Note to Chapter 4.

Dividing the world into the OPEC and the non-OPEC countries, our model expresses the number of years required to solve oil-caused deficits in terms of the relative economic growth rates of OPEC and non-OPEC countries as well as their respective income elasticity of demand for imports from each other. The model tells us that, given their respective income elasticity of demand for imports from each other, the faster the OPEC economy grows in relation to the non-OPEC economy, the sooner the oil deficits of non-OPEC countries will disappear. It also shows that a reduction in non-OPEC's income elasticity of demand for oil imports from the OPEC countries will accelerate the adjustment toward payments equilibrium for non-OPEC countries.

For the world economy, therefore, our model suggests two kinds of adjustment policies in order to ameliorate or overcome the current oil-induced payments imbalances: (1) raising the economic growth rate of OPEC countries relative to the growth rate of non-OPEC economies; and (2) economizing in oil consumption and developing oil substitutes by oil-importing countries. It should be noted, however, that our model does not necessarily suggest deflationary policies in non-OPEC countries;

it simply requires that the differential in the economic growth rates between OPEC and non-OPEC countries be increased as much as possible.

Technical Note to Chapter 4

In this note we will develop the dynamic adjustment model as an alternative to both the elasticity and the absorption approaches, first focusing our attention on the oil-caused trade imbalances and then generalizing the model to other payments imbalances. First of all, let us divide the world into OPEC and non-OPEC countries. Secondly, let us assume that the economies of OPEC countries as a group are to develop at a constant growth rate of g, while those of oil-importing non-OPEC countries as a group are to develop at a constant growth rate of h. Then, each group's combined GNP in the year t will be expressed respectively, as follows:

$$Y_{pt} = Y_{po} e^{gt} \quad \ldots\ldots\ldots\ldots \quad (1)$$

$$Y_{ct} = Y_{co} e^{ht} \quad \ldots\ldots\ldots\ldots \quad (2)$$

where Y_{pt} and Y_{ct} refer respectively to the combined GNPs of OPEC countries and oil-importing non-OPEC countries in the year t, and Y_{po} and Y_{co} refer to their respective GNPs in the base year when $t = 0$, assuming no change in the terms of trade. We will further assume that OPEC's oil and other exports to non-OPEC countries will grow at a constant rate of r, and that OPEC's imports from non-OPEC will grow at a constant rate of s. If m is non-OPEC's income elasticity of demand for imports from OPEC countries, and n is OPEC's income elasticity of demand for imports from non-OPEC countries,[12] m and n can be expressed by the growth rates of both exports (or imports) and the GNP, as follows:

$$m = \frac{\Delta M_c}{M_c} \div \frac{\Delta Y_c}{Y_c} = \frac{\Delta X_p}{X_p} \div \frac{\Delta Y_c}{Y_c} = r \div h \quad \ldots\ldots \quad (3)$$

$$n = \frac{\Delta M_p}{M_p} \div \frac{\Delta Y_p}{Y_p} = s \div g \quad \ldots\ldots\ldots\ldots\ldots\ldots \quad (4)$$

$$\therefore\ r = mh \quad \ldots\ldots\ldots\ldots\ldots\ldots \quad (3')$$

$$s = ng \quad \ldots\ldots\ldots\ldots\ldots\ldots \quad (4')$$

where M_c and M_p refer respectively to imports of non-OPEC countries and OPEC countries, and X_c and X_p refer to their respective exports. We can also express OPEC's exports to, and imports from, non-OPEC countries in the year t as follows:

$$X_{pt}\ (\equiv M_{ct}) = X_{po}e^{rt} = X_{po}e^{mht} \quad \ldots\ldots. \quad (5)$$

$$M_{pt}\ (\equiv X_{ct}) = M_{po}e^{st} = M_{po}e^{ngt} \quad \ldots\ldots. \quad (6)$$

where the additional subscripts t and o refer respectively to the year t and the base year. Since we ignore transportation and insurance costs, $X_p = M_c$ and $M_p = X_c$. The condition of trade balance in the year t between OPEC and non-OPEC countries is:

$$X_{pt} = M_{pt} \quad \ldots\ldots\ldots\ldots.. \quad (7)$$

By substituting Equation (5) and (6) into Equation (7), we have Equation (8), as follows:

$$X_{po}e^{mht} = M_{po}e^{ngt} \quad \ldots\ldots\ldots\ldots \quad (8)$$

Solving Equation (8) in terms of t, we have Equation (9), as follows:

$$\frac{e^{ngt}}{e^{mht}} = \frac{X_{po}}{M_{po}}$$

$$e^{(ng-mh)t} = \frac{X_{po}}{M_{po}}$$

$$(ng-mh)t = \ln \frac{X_{po}}{M_{po}}$$

$$\therefore\ t = \frac{1}{ng-mh} \ln \frac{X_{po}}{M_{po}} \quad \ldots\ldots\ldots \quad (9)$$

Equation (9) indicates that t (the number of years from a base year, which would be required to correct the present oil deficits) is determined primarily by two variables: g (the growth rate of OPEC's economy), and h (the growth rate of oil-importing non-OPEC's economy). The economic structures of OPEC as well as non-OPEC countries may largely predetermine both m (non-OPEC's income elasticity of demand for imports from the OPEC members, primarily oil) and n (OPEC's income elasticity of demand for imports from non-OPEC countries, primarily food and industrial products). The ratio X_{po} / M_{po} is also predetermined in the base year automatically. Equation (9) shows that, given m and n, the faster the OPEC economy grows in relation to the non-OPEC economy [i.e., the bigger (ng-mh)], the sooner will the oil deficits of non-OPEC countries disappear (i.e, the smaller t). To a certain extent, however, it may be possible to reduce m (non-OPEC's income elasticity of demand for oil imports from OPEC countries) by energy conservation measures and development of substitute energy sources. Equation (9) shows that a decline in m does reduce t.

Admittedly, the concepts of m and n pose the problem of aggregation. The income elasticity of demand for imports has traditionally been applied to a single country, not a group of countries as in our case, in which the elasticity of OPEC, as well as non-OPEC, countries is utilized for the group of countries. Imports of certain countries within a group, such as Iran and Saudi Arabia in the OPEC group, may be heavily influenced by non-income factors such as the Shah of Iran's wish to buy military goods, etc. On the other hand, the imports of other OPEC countries, such as Nigeria and Indonesia, may depend significantly on their income rather than on non-income factors. Therefore, in aggregating the income elasticity for a group of countries which have diverse economic systems, the concept of income elasticity should be used with special care. The best way to aggregate the income elasticities of different countries for a group-wide income elasticity may be to calculate the weighted average of the country elasticities, the weight being determined by each country's share of the group's total imports.

For the world economy, therefore, our model suggests two kinds of adjustment policies to ameliorate or overcome the current oil-induced payments imbalances: (1) to raise the economic growth rate of OPEC countries relative to the growth rate of non-OPEC economies; and (2) to economize oil consumption and develop oil substitutes in oil-importing countries. It should be noted, however, that our model does not necessarily suggest the adoption of deflationary policies in non-OPEC countries; it simply requires that the differential in the economic growth rates between OPEC and non-OPEC countries be increased as much as possible. It will be very instructive to measure the approximate magnitudes of the above variables in order to estimate the degree of difficulty facing the world economy in solving the problem of oil deficits.

By using 1974 as the base year (t = 0) and by replacing the GNP figure by the GNI (gross national income), we can claculate the magnitude of t. An estimate by the World Bank of OPEC's combined GNI in 1974 (excluding Ecuador and Gabon hereinafter, following the World Bank's practice) is $114 billion in 1967-69 U.S. dollars, while the Bank's projection of OPEC members' annual growth rate of real GNI is 7.1 percent for the period 1974-80.[13] Therefore, we will assume that:

$$Y_{po} = \$114 \text{ billion}$$

$$g = 7.1\% \text{ per annum}$$

Similarly, the World Bank's estimates for the rest of the world (excluding Communist countries) are as follows:

		OECD Countries	Non-oil LDCs	Total
1974 GNI in 1967-69 US $bil.	:	$2,133	$450	$2,583
Annual growth rate for 1974-80	:	4.6%	4.4%	4.6%

Therefore, we have:

$$Y_{co} = \$2,583 \text{ billion}$$

$$h = 4.6\% \text{ per annum}$$

According to Morgan Guaranty estimates, in 1974 OPEC countries' oil and non-oil exports amounted to $112 billion, and their combined imports of goods and services were $50 billion.[14] Thus, we have X_{po} = $112 billion and M_{po} = $50 billion. There have been no estimates made of either OPEC's income elasticity of demand for imports (n) or non-OPEC's income elasticity of demand for oil imports from OPEC countries (m). Therefore, we have to be satisfied with rough estimates. According to the World Bank's estimates, OPEC's imports are to grow at an average annual rate of 24.5 percent per annum for 1975-80.[15] Since OPEC economies are estimated to grow at 7.1 percent per annum for the same period, we may use a provisional figure of 3.5 as OPEC's income elasticity of demand for imports ($n = 24.5 \div 7.1 = 3.5$). Income elasticities of major Western countries have been calculated by Houthakker and Magee from data for 1951 through 1966.[16] According to their calculations, fifteen Western industrial countries had an average income elasticity of 1.3 for their combined imports, which included oil as well as non-oil imports. Due to lack of a better figure, we will first use 1.3 as the income elasticity of demand for imports by non-OPEC countries from OPEC countries.

How many years, then, would it take to achieve oil-trade balance by non-OPEC countries? From Equation (9),

$$t = \frac{1}{ng-mh} \ln \frac{X_{po}}{M_{po}}$$

With g = 7.1%, h= 4.6%, m = 1.3, n = 3,5, Xpo = $112 billion and Mpo = $50 billion,

$$t = \frac{1}{3.5 \times 0.071 - 1.3 \times 0.046} \ln \frac{112}{50}$$

$$= \frac{1}{0.249 - 0.060} \ln 2.24$$

$$= 5.3 \times 0.81$$

$$= 4.3, \text{ or about 4 years.}$$

This means that non-OPEC countries will eliminate oil deficits by the year 1978. We know that $X_{po}/M_{po} = 112/50 = 2.24$ was already predetermined in 1974, so policy makers can only influence $1/ng-mh$ to affect t. Let us now assume that non-OPEC countries reduce their income elasticity of demand for oil imports to 1 or even 0.5. With m = 1, oil imports will grow at the same rate as non-OPEC economies, while oil imports grow only half as fast as the non-OPEC economies with m = 0.5.

(1) With m = 1 and n = 3.5,

$$t = \frac{1}{3.5 \times 0.071 - 0.046} \ln 2.24$$
$$= 4.93 \times 0.81$$
$$= 3.99, \text{ or about 4 years.}$$

(2) With m = 0.5 and n = 3.5,

$$t = \frac{1}{3.5 \times 0.071 - 0.5 \times 0.046} \ln 2.24$$
$$= 4.42 \times 0.81$$
$$= 3.6 \text{ or, again, about 4 years.}$$

The above results show that oil conservation and/or development of oil substitutes may not be as effective as has been generally believed in restoring trade balance between OPEC and non-OPEC countries. Even if m is reduced to 0.5 from 1.3, t remains about the same. Let us now see the effect of a change in n, OPEC's income elasticity of demand for imports from non-OPEC countries, while leaving m = 1.3 unchanged.

The results are as follows:

n :	1.5	2	2.5	3	4
t :	17.4	9.9	6.9	5.3	3.6

The above table shows that when n is reduced from the original figure of 3.5 to 2 or 1.5, the required adjustment period increases rather substantially. For example, with n = 2 the required adjustment period climbs to 10 years, and it is further extended to 17 years with n = 1.5. Let us next see what happens when we increase the growth-rate differential between OPEC and non-OPEC economies. First, we will *decrease* the economic growth rate of non-OPEC countries by one percent from h = 4.6% to h = 3.6%, while leaving the growth rate of OPEC countries unchanged at g = 7.1%.

With m = 1.3, n = 3.5, g = 7.1% and h = 3.6%,

$$t = \frac{1}{3.5 \times 0.071 - 1.3 \times 0.036} \ln 2.24$$

$$= 4 \text{ years.}$$

Thus, there is almost no change in the required adjustment period, even if the annual growth rate of non-OPEC countries is reduced by one percent. Let us now *increase* the growth rate of OPEC countries by one percent from g = 7.1% to g = 8.1%, while leaving h = 4.6% unchanged. In this case, we find the required adjustment period reduced from 4 years to 3.6 years. Thus, we find that, in shortening the adjustment period, an expansionary policy to *increase* the growth rate of OPEC economies by one percent is more effective than a deflationary policy to *reduce* the growth rate of non-OPEC economies by the same one percent.

This result can be seen more clearly, if we use n = 2 instead of n = 3.5. In fact, our original assumption of n = 3.5 may be too high, considering m has been assumed to be only 1.3. It appears more realistic to have OPEC's income elasticity of demand for imports as 2 i.e., n = 2, rather than n = 3.5. Using n = 2 instead of n = 3,5, let us recalculate t for the two cases, when h is reduced from 4.6% to 3.6%, and when g is increased from 7.1% to 8.1%.

(1) When h is reduced from 4.6% to 3.6%, with m = 1.3, n = 2 and g = 7.1%: t is reduced from 9.9 years to 8.5 years, or by 14%.
(2) When g is increased from 7.1% to 8.1%, with m = 1.3, n = 2 and h = 4.6%: t is reduced from 9.9 years to 7.9 years, or by 20%.

How large will the payments surpluses accumulated by the OPEC countries be by the time OPEC's current-account surpluses finally disappear? Until the year k when $X_{pt} = M_{pt}$ the trade surpluses being accumulated annually by the OPEC countries as a group will be:

$$S_{pt} \; (\equiv D_{ct}) = X_{pt} - M_{pt} \quad \ldots\ldots\ldots\ldots \quad (10)$$

where S_{pt} is the annual OPEC trade surplus for the year t, identical with D_{ct}, which is the annual trade deficits of non-OPEC countries in the year t. Substituting Equations (5) and (6) into Equation (10),

$$S_{pt} = X_{po} e^{mht} - M_{po} e^{ngt} \quad \ldots\ldots\ldots\ldots\ldots\ldots \quad (11)$$

By integrating the annual increases in OPEC surpluses until the year k, when it is assumed that trade balance will exist between OPEC and non-OPEC countries, we can obtain the total accumulation of OPEC trade surpluses, Tp, from the base year when t = 0 until the year when t = k.

$$T_p = \int_0^k S_{pt}\,dt = \int_0^k (X_{po} e^{mht} - M_{po} e^{ngt})dt$$

$$= (\frac{1}{mh} X_{po} e^{mhk} - \frac{1}{ng} M_{po} e^{ngk}) + (\frac{1}{ng} M_{po} - \frac{1}{mh} X_{po}).$$

From Equation (8), however,

$$X_{po} e^{mht} = M_{po} e^{ngt}, \text{ when } t = k.$$

$$\therefore T_p = X_{po} e^{mhk} (\frac{1}{mh} - \frac{1}{ng}) + (\frac{1}{ng} M_{po} - \frac{1}{mh} X_{po}) \quad \ldots\ldots \quad (12)$$

GENERALIZED DYNAMIC ADJUSTMENT MODEL

The dynamic adjustment model which we have described so far in connection with the oil-induced payments imbalance between OPEC and non-OPEC countries can be easily generalized to the balance-of-payments adjustment problems of any country vis-à-vis the rest of the world. Instead of dividing the world into OPEC countries and non-OPEC oil-importing countries, we can think of a specific country A and the rest of the world W. In order to evaluate the specific trade-off involved for country A between economic growth and external-trade balance, we can replace the OPEC countries by country A, and the non-OPEC oil-importing countries by the rest of the world W. Then Equation (9) will be changed as follows:

$$t = \frac{1}{ng - mh} \ln \frac{X_{ao}}{M_{ao}} \quad \ldots\ldots\ldots\ldots \quad (13)$$

where all other symbols are used in the same way as in our dynamic adjustment model for oil-importing countries. Similarly, from Equation (12) we can calculate the amount of country A's accumulated surpluses (if A has a trade-surplus problem) or accumulated deficits (if A has a trade-deficits problem) by the time when A will have completed its trade adjustment, as follows:

$$T_a = X_{ao} e^{mhk} \left(\frac{1}{mh} - \frac{1}{ng}\right) + \left(\frac{1}{ng} M_{ao} - \frac{1}{mh} X_{ao}\right) \quad \ldots\ldots \quad (14)$$

If Country A has a trade-deficit problem it may want to solve this problem by the year k, so that T_a equals A's present foreign-exchange reserves, R_o. At the same time, we may calculate A's required economic growth rate, g, with a given h (the economic growth rate of the rest of the world), so that $T_a = R_o$ can be satisfied. From Equation (13), we have:

$$t = \frac{1}{ng - mh} \ln \frac{X_{ao}}{M_{ao}}$$

Since k is the year when country A obtains external trade balance, t = k. Thus, we can rewrite Equation (13) as follows:

$$k = \frac{1}{ng - mh} \ln \frac{X_{ao}}{M_{ao}} \quad \ldots\ldots\ldots\ldots \quad (15)$$

Similarly, we can rewrite Equation (14) by using the condition $T_a = R_o$, as follows:

$$R_o = X_{ao} e^{mhk} \left(\frac{1}{mh} - \frac{1}{ng}\right) + \left(\frac{1}{ng} M_{ao} - \frac{1}{mh} X_{ao}\right) \quad .... \quad (16)$$

We now have two equations, (15) and (16), and two unknowns, g and k. Consequently, Country A can simultaneously calculate the required economic growth rate (g) and the number of years necessary to restore external trade balance (k), given A's income elasticity of demand for imports (n) and the income elasticity of the rest of the world for demand for A's exports (m), as well as the economic growth rate of the rest of the world (h). Viewed in this light, the great controversy over the adequacy or inadequacy of international reserves should only be solved in a dynamic context, by considering a country's trade-off function between economic growth and external balance. A country's reserves will be considered "adequate", if and only if $R_o = T_a$ with a given domestic economic growth rate, g, which the country wants to maintain regardless of its external imbalance.

Our generalized dynamic adjustment model, by incorporating the time element in trade adjustment as well as the specific trade-off between the economic growth rate and the speed of payments adjustment, goes a long way toward remedying the shortcomings of essentially static conventional adjustment theories. Earlier in this chapter, we discussed why conventional adjustment theories such as the elasticity and absorption approaches are patently inadequate in prescribing the necessary adjustment process involved in the current oil-induced payments imbalances.

NOTES

[1] *World Financial Markets,* Morgan Guaranty Trust Company of New York, January 21, 1975, pp. 7-8.

[2] Gottfried Haberler, "The Market for Foreign Exchange and the Stability of the Balance of Payments: A Theoretical Analysis," *Kyklos,* Vol. III, 1949, pp. 197-199.

[3]Sidney Alexander, "Effects of a Devaluation on a Trade Balance," *IMF Staff Papers,* April 1952.

[4]Harry Johnson, "Towards a General Theory of the Balance of Payments," in Caves and Johnson, eds., *Readings in International Economics* (Homewood Ill.: Richard D. Irwin, 1968), p. 375.

[5]John Hein, "Oil Money and World Payments: The Billion Dollar Unsure Thing," *The Conference Board Record,* September 1974, p. 8.

[6]Henry Wallich, "Longer-run Aspects of International Monetary Reform," *Aussenwirtschaft,* September 1974, p. 340.

[7]J. R. Hicks, "An Inaugural Lecture," *Oxford Economic Papers,* June 1953, pp. 117-135.

[8]H. G. Johnson, "Economic Development and International Trade," *Nationaloekonomisk Tidsskrift,* vols. 5-6, 1959; and J. Bhagwati, "Immiserizing Growth: A Geometric Note," *Review of Economic Studies,* June 1958.

[9]Bo Soedersten, *A Study of Economic Growth and International Trade* (Stockholm: Almqvist & Wiksell, 1964). See also Bo Soedersten, *International Economics* (New York: Harper & Row, 1970), Chapter 9.

[10]Richard Caves and Ronald Jones, *World Trade and Payments* (Boston: Little, Brown and Company, 1973), Chapter 25 (The Effects of Growth on Trade), pp. 523-44; and Robert M. Stern, *The Balance of Payments: Theory and Economic Policy* (Chicago: Aldine Publishing Company, 1973), Chapter 11 (Economic Growth and the Balance of Payments), pp. 363-86.

[11]Robert M. Stern, *Ibid.*, pp. 364-75.

[12]For explanations and empirical estimates of income elasticities, see H. S. Houthakker and Stephen P. Magee, "Income and Price Elasticities in World Trade," *The Review of Economics and Statistics,* May 1969.

[13]*Prospects for Developing Countries: 1976-1980.* International Bank for Reconstruction and Development (Washington, D.C., July 8, 1975), Annex Table 1.

[14]*World Financial Markets,* Morgan Guaranty Trust Company of New York, October 20, 1975.

[15]*Prospects for Developing Countries, op. cit.,* Annex Table 7.

[16]H. S. Houthakker and Stephen P. Magee, *op. cit.*

Chapter 5

Effects of Oil Payments on Real Income

Sharp increases in oil prices have had a significant effect on the real incomes of both oil-exporting and oil-importing countries. According to one estimate, the quintupling of oil prices since late 1973 has resulted in a 2 percent reduction in the GNP of the OECD countries, a 3 percent GNP reduction on the part of the non-OPEC developing world, and a doubling of the GNP of the OPEC countries.[1] Real incomes are adversely affected for oil-importing countries by three factors related to the oil crisis. First, the terms of trade have worsened for oil-importing countries, with an attendant transfer of real income from oil-importing countries to oil-exporting countries. Second, increase in oil prices has an excise-tax effect on consumption, depressing real income through a reduction in the income multiplier. Third, the worsened trade balances of oil-importing countries tend to encourage the governments of these countries to adopt deflationary policies to reduce their payments deficits, thereby inducing slower economic growth than would be the case without the oil crisis. Apart from these three negative effects on the absolute level of real incomes of oil-importing countries as a whole, there is also a distribution effect on real incomes among residents of oil-consuming countries due to the increased cost of energy. These various real-income effects are discussed in the following sections.

TERMS-OF-TRADE EFFECT

Among the several different measures of the terms of trade, the most relevant one for the purpose of measuring the effect of oil-price increases on real income is the income terms of trade. It consists of the index of the value of exports divided by the index of import prices, as follows:

$$\text{Income terms of trade} = \frac{P_x Q_x}{P_m},$$

where P and Q refer to the prices and quanties for imports (m) or exports (x). If oil-price increases have not affected both the export prices and the physical quantities of exports of oil-importing countries, the income terms of trade of oil-importing countries may have deteriorated, as shown in the following formula:

$$\frac{P_x^o Q_x^o}{P_m'} < \frac{P_x^o Q_x^o}{P_m^o} \quad \ldots\ldots\ldots\ldots (1),$$

because $P_m' > P_m^o$, where P_m^o, P_x^o, and Q_x^o respectively refer to the original prices of imports and exports and the quantity of the exports, and P_m' refers to the changed import prices due to oil-price increases. Furthermore, the trade balance deteriorates for oil-importing countries if the volume of imports remains the same, because:

$$(P_x^o Q_x^o - P_m^o Q_m^o) > (P_x^o Q_x^o - P_m' Q_m^o) \quad \ldots\ldots\ldots \quad (2)$$

This means that, *ceteris paribus,* the real income of oil-importing countries has declined, which can be explained by using a simple national income model,

$$Y = C + I + (X - M) \quad \ldots\ldots\ldots\ldots \quad (3)$$

By differentiating the above equation to find the total differential, dY;

$$dY = dC + dI + (dX - dM) \quad \ldots\ldots\ldots\ldots \quad (4)$$

Since dC and dI are zero due the *ceteris paribus* assumption,

$$dY = dX - dM \quad \ldots\ldots\ldots\ldots\ldots\ldots \quad (5)$$

Earlier, we asumed no change in the prices and quantity of exports by oil-importing countries, so dX = 0. However, $dM = P_m' Q_m^o - P_m^o Q_m^o > 0$ due to oil-price increases as well as the assumption of no price-increase effect on the imports quantity.

$$\therefore dY = -dM < 0 \quad \ldots\ldots\ldots\ldots \quad (6)$$

Thus, Y has declined absolutely due to oil-price increases in this case. In reality, an increase in oil prices also affects consumption (C) and investment (I), lowering C and presumably raising I a bit; in other words, we may have $dC<0$ and $dI>0$. Therefore, the decrease in consumption may be offset by the increase in investment, and, in the end, the result can be similar to the one with our earlier *ceteris paribus* assumption of $dC = 0$ and $dI = 0$.

The income terms of trade is actually the net barter terms of trade times the index of the quantity of exports. The net barter terms of trade, also known as the commodity or merchandise terms of trade,is the most widely used terms-of-trade measure. It is simply an index of export prices divided by an index of import prices with the quotient expressed as a percentage, as follows:

$$\text{Net barter terms of trade} = \frac{P_x}{P_m} \times 100.$$

A decline in the net barter terms of trade means that a given volume of exports will exchange for a smaller volume of imports than before. But the index does not indicate what has happened to the physical volume of goods traded. The net barter terms of trade declines when the prices of imports rise more (or decline less) than the prices of exports. In such a case, however, it is also likely that the volume of imports will decline and the volume of exports will rise, the extent of the decline and the rise being dependent on the price elasticities of imports and exports.

Unlike the net barter terms of trade, the income terms of trade expresses the change in the physical volume of exports accompanying the change in the ratio of export prices to import prices. Still, the income terms of trade does not consider the change in the import volume; hence the index does not reflect the total changes in the trade balance due to the change in the ratio of export prices to import prices. Furthermore, the income terms of trade does not show what happens to real absorption possibility if the economies of importing countries are fully employed. The income terms of trade and the net barter terms of trade can also move in opposite directions. If import prices remain constant but export prices fall by 5 percent, the net barter terms of trade deteriorate from 100 to 95. If, however, the quantity of exports increases by 10 percent due to the 5-percent fall in export prices, the income terms improves from 100 to 104.5, as follows:

$$\frac{P_x}{P_m} \cdot Q_x = \frac{95}{100} \times 110 = 104.5 \ .$$

There are three other measures of the terms of trade. The gross barter terms of trade is the index of the quantity of exports divided by the index of the quantity of imports, with the quotient expressed as a percentage ($Q_x/Q_m \times 100$). The shortcoming of this index is that it considers neither export prices nor import prices. Two other concepts—the single and double factoral terms of trade—relate to the exchange of productive factors embodied in traded goods. The single factoral terms of trade consists of the net barter terms of trade adjusted for productivity changes in the production of exports. An adverse change in a country's net barter terms of trade suggests that the country is worse off than in the base year, because it has to export more goods to pay for a given quantity of imports.

In reality, however, the decline in export prices may have resulted from technological progress that lowered production costs. If the decline in production costs is more that proportionate to the decline in the terms of trade, the country is better off even though its net barter terms of trade has deteriorated, because the country obtains more imports per unit of productive factors employed in producing exports than it did in the base year. The double factoral terms of trade considers productivity changes in both exports and imports. However, recent oil-price increases did not result from relative productivity changes, but from unilateral cartel-type decisions by oil-exporting countries. Hence, factoral terms of trade are also not suitable for evaluating the effect of oil-price increases on real incomes.

So far, we have seen that most measures of the terms of trade consider either prices (the net barter, single and double factoral terms of trade) or quantities, (the gross barter terms of trade). Only the income terms of trade includes both the prices of exports and imports and the quantity of exports. But even the income terms of trade does not consider changes in the quantity of imports, and we had to employ strict *ceteris paribus* assumptions regarding the export and import quantities in our earlier calculation of the oil-price effect on real incomes. In fact, with the *ceteris paribus* assumptions governing the quantities, the income terms of trade was used in the same way as the net barter terms of trade. In other words, the effect of oil-price increases on export and import quantities was completely ignored.

In order to evaluate the effect of price changes on real incomes while at the same time taking into consideration their effect on the volume of exports as well as imports, we may need a new measure of the terms of trade incorporating not only changes in the prices of both exports and imports, but also changes in the quantities of both exports and imports. This we will call the total terms of trade:

$$\text{Total terms of trade: } \frac{P_x Q_x}{P_m Q_m} \times 100$$

But the above index is simply another way of expressing a country's trade balance in terms of a ratio rather than a difference. Some of the changes in the export and import quantities may not result from price changes, but from changes in tastes, income, population, etc. On the other hand, oil-price increases have caused changes in both prices and quantities of exports and imports by oil-importing countries. First of all, the oil-price increase led to a rise in the index of import prices of oil-importing countries. It also contributed to a decline in the volume of oil imports, a decline further exacerbated by the widespread economic recession in the West, which in turn was partly caused by the oil-price increases. Oil-price increases also contributed to world inflation, which has been responsible for the higher prices of exports from oil-importing countries. Finally, increased oil revenues have encouraged more imports by oil-exporting countries from oil-importing countries. In summary, oil-price increases have caused changes in P_x, P_m, Q_x, and Q_m for oil-importing countries.

Let us refer to the changes in P_x, P_m, Q_x, and Q_m, induced directly or indirectly by the oil-price increases, as dP_x, dP_m, dQ_x, and dQ_m, respectively. Then we can calculate the amount of real-income transfer (dT) from oil-importing to oil-exporting countries due to the oil-price increases. As noted earlier, oil-price increases were due not to productivity changes, but to the unilateral OPEC decision to raise prices. Subsequent changes in the prices of exports from oil-importing countries were also due not to productivity changes, but to worldwide inflation, partly caused by the oil-price hikes. Hence, the amount of real-income transfer from oil-importing to oil-exporting countries is equivalent to the amount of real-income decline in oil-importing countries due to the worsened terms of trade. In other words, if we ignore the secondary income effect resulting from changes in both consumption and investment due to oil-price hikes, the amount of real-income increase for oil-exporting countries is equal to the amount of real-income decrease for oil-importing countries caused by the worsened terms of trade for the latter. From Equation (5), we calcualted this real-income decline as:

$$dY = dX - dM \quad \ldots\ldots\ldots\ldots\ldots \quad (5)$$

Since the amount of real-income transfer (dT) is the negative of dY, we have:

$$dT \; (\equiv - dY) = dM - dX \quad \ldots\ldots\ldots\ldots \quad (6)$$

Oil-price increases have caused changes in both the prices and quantities of exports and imports by dP_x, dP_m, dQ_x and dQ_m, respectively. Therefore, oil-price induced changes in M (the value of imports) and X (the value of exports) are:

$$dM = (Q_m + dQ_m) \cdot (P_m + dP_m) - Q_m P_m \quad \ldots\ldots\ldots \quad (7)$$

$$dX = (Q_x + dQ_x) \cdot (P_x + dP_x) - Q_x P_x \quad \ldots\ldots\ldots \quad (8)$$

Substituting Equations (7) and (8) into Equation (6),

$$dT = dM - dX$$

$$= (Q_m \cdot dP_m + P_m \cdot dQ_m + dQ_m \cdot dP_m)$$

$$- (Q_x \cdot dP_x + P_x \cdot dQ_x + dQ_x \cdot dP_x) \quad \ldots\ldots\ldots \quad (9)$$

Some economists have calculated dT as the net increase in the value of oil imports. For example, Robert Ulin, using the net increase in oil-import bills of $60 billion in 1974 as the amount of real-income transfer to OPEC countries, says that it is "less than 2 percent of the combined ($3 trillion-plus) GNP of the OECD countries in 1974."[2] Robert Triffin also considers the amount of real-income transfer to OPEC countries as equivalent to the increase in net oil-imports bills. Triffin notes: "Measured in GNP terms, the real transfers entailed by the increase in oil prices, even on the most pessimistic assumptions recently advanced, would be of the order of one to two percent a year for the OECD as a whole (with a maximum of 6 percent for Ireland and less than 4 percent for Italy)."[3] Triffin presents the following table (Table 5.1) to indicate the magnitudes of real-income transfers from oil-importing countries to OPEC member countries.

Table 5.1
Real Transfers to OPEC from OECD Countries in 1974

		Increase in net oil import bills in 1974 ($ billions)	% of GNP
I.	EEC	25.6	2.4%
	France	6.3	2.4
	Germany	6.2	1.8
	Italy	5.0	3.6
	Netherlands	0.5	0.8
	Belg - Lux.	1.3	2.8
	U.K.	5.2	3.0
	Ireland	0.4	6.0
	Denmark	0.7	2.6
II.	Other Western Europe	6.1	2.4%
III.	Other OECD	27.3	1.7%
	U.S.	14.7	1.1
	Canada	(0.2)	(0.2)
	Japan	12.0	2.9
	Australia) New Zealand)	0.8	1.1
IV.	Total OECD	59.0	1.8%

Source: Robert Triffin, *ibid.*, p. 24.

In calculating the amount of real-income transfer to OPEC members, both Ulin and Triffin ignore the effect of oil-price increases on both prices and quantities of exports from oil-importing countries to OPEC countries, that is, they assume $dP_X = 0$ and $dQ_X = 0$. Hence, the amount of real-income transfer (dT) is considered by them as equal to the increase in net oil-import bills for oil-importing countries. Consequently, the real transfer as defined by both Ulin and Triffin may be calculated as follows:

$$dM = (Q_m + dQ_m) \cdot (P_m + dP_m) - Q_m \cdot P_m \quad \ldots\ldots \quad (7')$$

$$\begin{aligned} dX &= (Q_x + dQ_x) \cdot (P_x + dP_x) - Q_x \cdot P_x \\ &= (Q_x + 0) \cdot (P_x + 0) - Q_x \cdot P_x = 0 \quad \ldots\ldots \quad (8') \end{aligned}$$

$$\begin{aligned} \therefore\ dT &= dM - dX \\ &= (Q_m + dQ_m) \cdot (P_m + dP_m) - Q_m \cdot P_m \quad \ldots\ldots \quad (9') \end{aligned}$$

But we have seen earlier that dP_X and dQ_X are not zero. Consequently, the amount of real-income transfer is actually less than the increase in net oil-import bills. In reality, however, it is difficult to measure dQ_m, dQ_X, and dP_X caused by dP_m (the increase in oil prices), because factors other than dP_m may also have influenced dQ_m, dQ_X, and dP_X. An alternative way to measure the real transfer may be to assume that both dQ_m and dQ_X are zero, and that dP_X (the increase in the prices of OPEC imports from the oil-importing countries directly and indirectly caused by the oil-price increase) is the same as the overall price increase (including both the oil-caused increase and other increases) of the OPEC imports, represented by dP_{XO}. In this case, we are no longer talking about the amount of real-income transfer from oil-importing to oil-exporting countries caused *only* by the recent oil-price upsurge. Instead, we are now considering the amount of real-income transfer caused by both the oil-price increase and other factors, such as the increase in the price of OPEC imports owing to food shortages, general price inflation, etc. The amount of real transfer in this case is calculated as follows:

$$dM = (Q_m + dQ_m) \cdot (P_m + dP_m) - Q_m \cdot P_m$$

$$= (Q_m + 0) \cdot (P_m + dP_m) - Q_m \cdot P_m$$

$$= Q_m \cdot dP_m \quad \ldots\ldots\ldots\ldots\ldots\ldots \quad (7'')$$

$$dX = (Q_x + dQ_x) \cdot (P_x + dP_{xo}) - Q_x \cdot P_x$$

$$= (Q_x + 0) \cdot (P_x + dP_{xo}) - Q_x \cdot P_x$$

$$= Q_x \cdot dP_{xo} \quad \ldots\ldots\ldots\ldots\ldots\ldots \quad (8'')$$

$$\therefore dT = dM - dX$$

$$= Q_m \cdot dP_m - Q_x \cdot dP_{xo} \quad \ldots\ldots\ldots\ldots \quad (9'')$$

Even though the real income of oil-importing countries has declined because of the oil-price hike, it is not clear whether a lowering of the living standard in terms of reduced real consumption has actually taken place in all oil-importing countries. In any case, it is necessary to distinguish the effect of the oil-price increase on real income from its effect on the standard of living.[4] If an oil-importing country borrows funds from abroad to pay for the incremental portion of oil payments caused by oil-price hikes, the lowering of the living standard may be deferred. In this case, the real burden of oil-price hikes will fall on future generations. Already, many oil-importing countries have borrowed rather heavily from abroad to pay for oil imports. To the extent that oil imports are financed by overseas borrowings, the effects of real-income decline on the living standard in oil-importing countries can be delayed, if additional savings realized through the increased retail price of oil are plowed back right away into the economy through an equivalent increase in investment.

EXCISE-TAX EFFECT ON REAL INCOME

Higher prices for oil have had a dampening effect on consumer spending analogous to that resulting from an increase in excise tax, the result in either case being a decrease in consumption and an increase in resources available for investment. When oil prices were sharply increased in late 1973, most industrialized countries were applying tight fiscal and monetary policies to combat excess demand and inflation. Insofar as the deflationary income effect of the oil-price increase was recognized, it was welcomed by some as a reinforecment of anti-inflation policies. In addition, it was pointed out that the adjustment to higher oil prices would require a considerable volume of new investment spending. Thus, it made sense to many observers that the oil "excise tax" should bring with it a shift from spending to saving.[5]

However, since the oil-price increase is analogous to a consumption tax rather than an income tax, its initial effect is to raise prices, creating an immediate inflation before the ultimate deflationary effect, from less consumption spending, is felt. Furthermore, it is essentially a tax imposed from outside rather than as part of a government plan designed to achieve particular national objectives. Therefore, the disposition of the "tax" proceeds is under the control of the OPEC countries, not of the oil-consuming countries. The inability of OPEC countries to spend all of their increased oil revenues either through consumption or investment causes an outward shift in the world savings schedule. The degree to which the additional savings are offset by an increase in investment depends on the shape of the savings and investment schedule. Under an extreme monetarist view, the "excise tax" effect of oil-price hikes would not affect the aggregate demand composed of consumption and investment if the oil deficit were exactly offset by capital inflows under a fixed exchange-rate system, since it would have no effect on money balances. Under the floating-rate regime, the oil deficit need not affect money balances at all, even if it is not offset by caital inflows. There might only be a shift in the distribution of aggregate expenditure toward more investment at the expense of consumption, since the outward shift in the world savings schedule would reduce interest rates and thus encourage investments.[6]

Under Keynesian assumptions, however, the decrease in consumption resulting from both higher oil prices and the concomitant reduction in real income is not fully offset by any increase in investment due to lower interest rates. In this case, the "exise tax" effect definitely reduces the aggregate demand, and, through the multiplier effect, leads to a still lower level of real income. However, the effect of the "excise tax" falls unevenly within the economy on energy-intensive industries and other industries. In the United States, for example, where over one-third of the oil consumed is used for automobiles, the automotive industry and other related industries (tourism, for example) are more severely affected than others. In other countries, the impact may come through higher costs of fertilizer or other energy-intensive imports. While industries using oil intensively may suffer more heavily than the rest of the economy, some industries producing non-oil energy sources, such as coal or nuclear power, benefit from the oil-price hikes.

The differential effect of oil-price hikes on the various sectors of a nation's economy has two important economic-policy implications. First, the industry,rather than the country, may be the appropriate unit for economic analysis of the effect of oil-price changes. Workers in the same industry in different countries are likely to be more similarly affected by the oil-price increases than are workers in different industries within the same country. The automotive industry is depressed throughout the world, while coal miners everywhere may enjoy an increase in real income, because the increase in oil price has raised the demand for oil substitutes. Second, expansionary macroeconomic policies, which are frequently recommended as remedies for the decline in real income resulting from deficient aggregate demand, may be less appropriate in dealing with the current problem caused by oil-price hikes. Rather, selective measures should be adopted to induce productive factors to move from the industries in which demand has fallen into the industries in which demand has risen as the result of oil-price increases.

Comparison to the "excise-tax effect" should not be pushed too far, however. An increase in excise tax rarely, if ever, creates large new investment outlets in industries producing close substitutes; the oil-price increase undoubtedly has done so. It has been estimated that the additional capital spending in the energy industries of eleven Western industrial countries is likely to amount to about $200 billion a year (in 1973 dollars) in the late 1970s.[7] This estimate, which was based on the stated intentions of the respective governments, implies an increase in total capital expenditure of over 50 percent, and an absolute increase in investment something like three or four times as large as the cumulative financial surpluses of the OPEC countries by 1980. Consequently, it has been argued that, while there may be a case for some reflationary measures in the short run in order to limit the deflationary impact of the oil-price increase that will otherwise occur before energy investment programs are implemented, the medium to long-run policy goal should be the adoption of additional measures of restraint in order to free resources for higher energy investment.[8]

DEFLATIONARY POLICY TO COPE WITH OIL DEFICITS

The oil-importing countries have to cope with the steep increase in their oil-import bills; most of these countries suffer big trade deficits due to the oil-price increases. One of the conventional methods of reducing trade deficits is to deflate the economy, in the belief that the deficit country exports less and imports more due to its over-heated economic activity. Since the increase in oil prices has generated an inflationary phenomenon, some oil-importing countries may take this development, along with the increased trade deficits, as an indication of economic over-heating, and may be tempted to resort to a deflationary policy to improve their trade balances. Such a policy, however, will lead to a further reduction in the real income of oil-importing countries. Unlike the terms-of-trade and "excise tax" effects, such a decline in real income is not an automatic consequence of the oil-price increases; it is the result of a deliberate (and perhaps misguided) policy adopted by an oil-importing country.

As we noted in the previous chapter, in the face of the very limited absorptive capacity of OPEC countries, the adoption of deflationary policies by some oil-importing countries may not significantly reduce the overall trade deficits of oil-importing countries vis-à-vis OPEC members. More likely, it will only lead to further deterioration of trade balances in other oil-importing countries, which will thus be forced to take counter-measures in the form of trade barriers, devaluation and/or their own deflationary policies.

* * *

In the preceding sections, we have studied three causes for decline in real income in individual oil-importing countires. Looking at these causes in the oil-importing countries as a whole, we can also foresee the likely appearance of cumulative international processes tending to reinforce the deflationary impact on these countries. The rise in oil prices simultaneously exerts deflationary influences on all oil-importing countries, and such a worldwide synchronization of deflationary impact can lead to an amplification of its effects through international economic linkages. Thus, we have to consider the deflationary effect on a country from the international multiplier as well as from the national income multiplier.

An econometric study by A. N. R. Schwartz shows that such international linkage effects, generated by interdependent international economic relationships, are indeed substantial, reinforcing the initial national deflationary impact of oil-price rises on real income. For example, the "partial" study of the effects of oil-price hikes on West Germany, using a separate, unlinked country model, shows a cumulative decline in the West German GDP of 3.3 percent after four years, while the "simultaneous" study, including the secondary effects from international linkages, shows a cumulative decline of 7.1 percent after four years.[9] This result indicates that the current oil crisis is a truly global problem, which requires a global, rather than individual-national, solution to fight against the serious deflationary impact of oil-price hikes.

NOTES

[1]Roger D. Hansen "The Emerging Challenge: Global Distribution of Income and Economic Opportunity," in James W. Howe and the Staff of the Overseas Development Council, *The U.S. and World Development: Agenda for Action 1974* (New York: Praeger Publishers, 1974), p. 157.

[2]Robert P. Ulin, "Oil Payments and the Economy," *The Conference Board Record,* May 1975.

[3]Robert Triffin, "The Threat to World Economic Order: The World "Infession" and the Oil Explosion," (Paper prepared for the Japan-U.S. Assembly at the American Enterprise Institute, Washington, D. C., April 7 and 8, 1975), p. 15.

[4]For a different view, see Norman S. Fieleke, "Oil and International Payments," *New England Economic Review,* November/December 1974, pp. 30-32.

[5]Robert Solomon, Adviser to the Board of Governors of the Federal Reserve System, "Impact of Higher Oil Prices," (Address before the Fourth European Institutional Investor Conference, London, December 3, 1974), p. 2.

[6]Andrew Crockett and Duncan Ripley, "Sharing the Oil Deficit," *IMF Staff Papers,* July 1975, p. 287.

[7]T. M. Rybczynski, "Capital Requirements for Energy Development in the Main Industrial Countries," (Paper presented to the British-North American Association, April 1974).

[8]John Williamson, "The Impact (of the Oil Price Increase) on the International Economy," (Paper presented at a Brookings Institution Conference on November 18-20, 1974), p. 14

[9]A. N. R. Schwartz, *The Effects of the Rise in Oil Prices on the Economy of Industrialized Countries: A Methodological and Quantitative Investigation* (The Hague: Dutch Central Planning Bureau, 1974), p. 13.

Chapter 6
Oil Money and the Eurocurrency Market

The impact of oil-price increases has been pervasive in many sectors of the world economy. In this chapter, we will analyze the effect of oil money on the Eurocurrency market. When the steep oil price increases were imposed in 1973, it became clear that the Eurocurrency market would play a major role in intermediating the expected huge oil-money flows. As the foremost international financial market, the Eurocurrency market is well suited to this function in terms of size, flexibility, technical skills, and a frequently-demonstrated adaptability to new circumstances. This chapter explores in some detail the various economic implications arising from the Eurocurrency market's intermediary role for oil money.

Since the Eurocurrency market is closely interconnected with major domestic money markets, our analysis naturally involves many important issues affecting the world economy. We will discuss the effects of oil money on the supply of and demand for Eurocurrency funds in the first two sections, which will serve as a prelude to the following section, which presents an analysis of the larger question of the effect of oil money on the volume of world credit. Then, we will deal with the role of the Eurocurrency market as a recycling channel for oil money, as well as the structural changes in the Eurocurrency market necessitated by this new role. The chapter concludes with a section on the changes taking place in the Eurobond market, which is a long-term capital market maintaining symbiotic coexistence with the short-term Eurocurrency money market. Tables 6.1 and 6.2 present some of the basic statistics of the Eurocurrency market in order to familiarize us with the market.

Table 6.1
Growth of the Eurocurrency Market*
(in billions of U.S. dollars)

end of period	estimated market size Gross**	Net†	changes in amount Gross	Net	changes in % Gross	Net
1964	21	14				
1965	24	17	3	3	14	21
1966	29	21	5	4	21	24
1967	36	25	7	4	24	19
1968	50	35	14	10	39	40
1969	86	50	36	15	72	43
1970	113	65	27	15	31	30
1971	155	85	42	20	37	31
1972	203	110	48	25	31	29
1973	305	160	102	50	50	45
1974 Mar	339	175	34	15	-	-
Jun	366	185	27	10	-	-
Sep	354	190	-12	5	-	-
Dec	373	210	19	20	-	-
Year	-	-	68	50	22	31
1975 Mar	382	220	9	10	-	-
Jun	392	230	10	10	-	-
half-year	-	-	19	20	10‡	19‡
mid-Aug	395	230				

*Based on foreign currency liabilities to residents and nonresidents reported by banks in eight European countries covered by BIS data (Belgium-Luxembourg, France, Germany, Italy, Netherlands, Sweden, Switzerland, and the United Kingdom) and by banks in Canada, Japan, the Bahamas and Cayman Islands, Singapore, and Panama.

**All foreign currency liabilities to residents and nonresidents, banks and nonbanks.

†Excludes interbank redeposits within the reporting area.

‡Annual rate

Source: Morgan Guaranty Trust Company of New York.

SUPPLY OF OIL MONEY TO THE EUROCURRENCY MARKET

The Eurocurrency market offers many advantages to international investors. It is bigger than other money markets with the exception of the U.S. market, and thus is able to offer liquid

Table 6.2
Eurocurrency Market Centers

	Gross foreign currency liabilities			
	end of 1970		end of 1974	
	amount in $ billions	% of total	amount in $ billions	% of total
banks in:				
United Kingdom	50.4	44	153.5	41
France	9.5	8	33.7	9
Luxembourg	2.5	2	20.1	5
Italy	10.1	9	15.3	4
Netherlands	4.2	4	13.6	4
Belgium	4.9	4	13.4	4
Switzerland	6.1	5	11.9	3
Germany	2.9	3	7.7	2
Sweden	0.5	-	1.0	-
Subtotal, Europe	91.1	80	270.2	72
Bahamas and Caymans	7.6	7	40.0	11
Japan	5.5	5	30.7	8
Canada	8.7	8	16.8	5
Singapore	0.4	-	10.4	3
Panama	0.4	-	5.2	1
Subtotal, other	22.6	20	103.1	28
Total, all centers	113.7	100	373.3	100

Source: Morgan Guaranty Trust Company of New York.

investment instruments to holders of large pools of funds. The market is anonymous and politically neutral (or even apolitical), which is important to oil countries that may fear the expropriation of, or the imposition of exchange controls on, their investments in domestic national markets. Unlike most national markets, the Eurocurrency market pays interest without withholding tax, and the yield is generally higher than that available in domestic markets. Banks in the Eurocurrency market can pay higher yields to depositors than, say, banks in the United States, for the following reasons: the Eurobanks have no mandatory reserve requirements like the banks in the United States under Regulation D; the Eurobanks engage in wholesale banking, and thus are able to save on administrative costs and operate with narrower profit margins per loan dollar; and there is no Regulation Q, which regulates interest-rate ceilings on deposits in the United States.[1]

Because of these advantages, OPEC countries had traditionally kept the bulk of their foreign-exchange reserves in the Eurocurrency market even before the 1973 oil-price upsurge. At the end of 1972, for example, about two-thirds to three-quarters of OPEC members' foreign-exchange reserves were kept in the Eurocurrency market, with the exception of Venezuela, which held most of its exchange reserves in the United States.[2] Therefore, from the start there was a general expectation that a substantial portion of OPEC's surplus oil revenues would flow to the Eurocurrency market. Developments in 1974 did indeed fulfill this expectation to a large extent. As the following table shows, $21 to $23 billion of OPEC surplus revenues flowed to the Eurocurrency market in 1974, as estimated by four different institutions.

Table 6.3
Estimated Deployment of OPEC Surpluses in 1974 (billions of U.S. dollars)

	BIS	US Treasury	OECD	Bank of England
Eurocurrency market	23	21	21	23
Bank deposits & securities in US	12	11	11	11
Bank deposits & securities in UK	5	7	6	7
International official institutions	4	4	4	4
Others	26	17	15	11
Total	70	60	57	56

Source: *Oil-Related Payments Imbalances,* Bank for International Settlements, March 11, 1975; Statement of the U.S. Treasury Secretary before the U.S. Congress Joint Economic Committee, February 5, 1975; *Financial Market Trends,* OECD, Paris, February 17, 1975; and *Quarterly Bulletin,* Bank of England, March 1975.

In addition to the placement of funds by OPEC countries in the Eurocurrency market, there are two other oil-related factors which have affected the supply of funds to the Eurocurrency market. First, there has been an increase on the order of several billion dollars in Eurocurrency deposits by oil companies.[3] As the amount of oil payments increased almost four times in 1974, to $115 billion from $32 billion a year earlier, the transactions balances of oil companies have had to increase also. It is true that almost all Eurocurrency deposits are time deposits, and the portion of call and overnight Eurocurrency deposits held by nonbanks is quite small. Therefore, the principal means of payment in the Eurocurrency market is the demand deposit at banks located in the country of the currency involved. However, as far as Eurodollars are concerned, there has been an increasing preference among multinational companies (MNCs) to hold money balances in the form of Eurodollar deposits rather than in the form of demand deposits in the United States.

This tendency among MNCs to hold Eurodollar deposits as a substitute for an increasing portion of their transactions balances has been due to the change in cash management methods of corporate treasurers that began in the mid-sixties. As the opportunity cost of money in financial markets has increased over the past decade, corporate treasurers have become increasingly reluctant to keep demand deposits as precautionary or transactions balances. As a result, most corporate demand deposits now belong to obligatory "compensating balances," which have to be maintained as part of the agreement in drawing down bank loans. The proliferation of interest-carrying substitutes for demand deposits has aided corporate treasurers in holding down the level of demand deposits to a minimum. Companies can invest idle money balances in highly liquid but interest-carrying near-monies such as CDs, Treasury bills and bankers acceptances. For example, a well-known international institution keeps only a minimum amount of demand deposits, even though its liquid money balances amount to more than $4 billion. These liquid money balances are invested almost constantly in interest-yielding money-market instruments. The same is true with most other institutions. Thus, there has been a certain change in the character of demand deposits in the past decade. Economists in general have been slow to recognize this change and continue to consider demand deposits as the primary precautionary transactions balances.

The "repo market," where funds can be safely loaned at a certain interest rate pledged with high-quality securities for a period as short as overnight, also allows corporate treasurers to make use of idle money balances profitably with the assurance of almost the same liquidity as demand deposits. There are primarily two types of "repos," or repurchase contracts. CD repos are secured by CDs, while Treasury repos are secured by Treasury securities. Since CDs are more risky than Treasury securities, CD repo rates are usually higher than Treasury repo rates by about 10 basic points. On the other hand, the Treasury repo rate is closely tied to the Federal funds rate. Repos may be arranged for overnight or longer. A typical repo contract works as follows: X owns highly liquid securities but is short of cash. When X wants to raise cash without necessarily selling outright his security holdings, X will arrange a repo contract. In this case, X sells, say, $10 million worth of Treasury bills to Y, with a promise to repurchase them after a fixed date. It is a repo contract for X, while it is a *resale* contract for Y. The repurchase price promised by X is calculated on the basis of repo interest rate. While CDs, Treasury securities, and commercial papers are all used as collateral, most repo contracts are executed using Treasury securities. Therefore, X has two ways to raise money quickly: he can either sell his securities outright or repo them. X will engage in repo contracts when he expects the interest rate to decline, for he can sell his securities later at a higher price.

The decline in the use of demand deposits as transactions balances in the United States has spilled over to the Eurodollar market, where multinational corporations increasingly keep their idle balances in the form of Eurodollar deposits or London dollar CDs without significant loss of liquidity. Therefore, it is quite understandable that oil companies have increased Eurodollar deposits in the past year as part of their growing oil transactions balances, thus raising the supply of funds in the Eurocurrency market.

The second oil-related factor which influences the supply of funds in the Eurocurrency market is the investment behavior of oil-importing countries. There are three cases to be considered here. In the first, the oil-importing countries that had previously kept some of their reserves in the Eurocurrency market have withdrawn their deposits to pay for oil imports. Since not all oil payments came back to the Eurocurrency market, as was shown in Table 6.3, some of those withdrawn Eurodollars never returned to the market, thus decreasing the supply of funds in the Eurocurrency market.

In the second case, some oil-importing countries have borrowed in the Eurocurrency market to accumulate funds to finance future oil imports. Frequently, these countries have kept their Eurocurrency loan proceeds in the market until they were needed. As long as they keep the loan proceeds in the Eurocurrency market, their borrowing of Eurocurrencies has no effect on the total supply of funds in the market. Once the proceeds are withdrawn to pay for oil imports, the effect is the same as in the first case, that is, a small net decrease in the supply of funds in the Eurocurrency market.

In the third case, some oil-importing countries have borrowed funds in national markets and then transferred their loan proceeds to the Eurocurrency market for investments until needed for oil payments. In this case, there is a net increase in the supply of funds in the Eurocurrency market. The overall effect of the above three cases on the supply of funds in the Eurocurrency market is difficult to measure, but it appears to be slightly negative. According to the BIS estimate, the net size of the Eurocurrency market grew by $55 billion in 1974 to a total of about $210 billion, as compared to a growth of some $50 billion in 1973.[4] Since OPEC supplied $23 billion and oil companies several billion dollars to the Eurocurrency market, the total supply of surplus oil funds to the market may have been about $25 billion, disregarding the contribution of oil-importing countries. However, the growth of the Eurocurrency market due to oil money must have been higher than $25 billion because of the Eurocurrency multiplier effect, to be discussed in a later section. It appears safe to conclude, therefore, that the bulk of the $55-billion Eurocurrency growth in 1974 was due to oil money, either directly or indirectly.

CHANGES IN THE DEMAND FOR EUROCURRENCY FUNDS

Both the Bank of England and the Bank for Internationl Settlements report on the foreign-currency assets and liabilities of banks engaged in Eurocurrency business. According to the BIS data,[5] foreign-currency liabilities to the Middle East of reporting European banks increased by almost $20 billion in 1974—to $29.8 billion from $10.3 billion in 1973—while their assets to the region increased by only $1 billion, from $2.6 billion to $3.6 billion. Some may interpret these figures as an indication that there was a net supply of funds in 1974 by $19 billion from the Middle East (mostly oil countries) to the Eurocurrency market. Similarly, from the fact that foreign-currency assets to Eastern Europe increased by $2.4 billion in 1974 while foreign-currency liabilities to the area rose only $1.3 billion, some may be tempted to conclde that Eastern European countries were a net user of Eurocurrency funds by $1.1 billion, and so on. Based on this line of reasoning, the BIS Annual Report concludes: "All in all the Eurobanks seem to have lent more to oil-importing countries outside their own area than they themselves received in the form of direct deposits from the oil-exporting countries."[6]

However, such an interpretation of the sources and destination of Eurocurrency funds is based on the erroneous concept that a geographic classification of foreign-currency "liabilities" of banks located in Europe can show from which areas the funds have come, and that a geographic classification of the foreign-currency "assets" of these banks shows to which areas the funds have gone. Machlup explains that this error in interpretation come from the "fallacy of composition": the fallacy of assuming that what is true for one bank is true for all banks together.[7] For an individual bank, it holds true that the depositor is the source of funds, and the borrower of the bank's loan is identified by the bank's assets. For all banks together, however, a geographic classification of Eurocurrency liabilities cannot tell us where the Eurocurrencies have *originated*, but only where they have *come to rest*. Similarly, the assets side of Eurobanks may not tell us where the credit end-users have come from, but only where the Eurocurrency funds are placed at the time, either temporarily or permanently.

A more reliable guide to the geographic location of Eurocurrency end-users is the data on the actual borrowers of syndicated Eurocurrency loans, known as Eurocredits. Though perhaps a quarter or more of the total Eurocredits are not publicized, the data on the publicized Eurocredits can give us a fair indication of the nature of the borrowers and the amounts involved.

Table 6.4
Publicly Announced Eurocredits, 1970-74 (millions of U.S. dollars equivalents)

Borrowers from:	1970	1971	1972	1973	1974	Total 1970-74
Developed countries	4,226	2,406	3,557	11,233	16,925	38,347
Developing countries	300	1,001	1,773	5,993	8,914	17,981
OPEC countries	146	350	961	3,013	773	5,243
Socialist countries	48	66	499	955	1,017	2,585
Int'l organizations	-	140	70	829	995	2,034
Total	4,720	3,963	6,860	22,023	28,624	66,190

Source: International Bank for Reconstruction and Development.

The total Eurocredit volume of $28,624 million for 1974 was an increase of 30 percent over the 1973 volume of $22,023 million and six times the 1970 volume of $4,720 million. Most of the growth, however, took place in the first half of 1974, while the volume for the second half was less than that for the comparable period of 1973.

Table 6.5
Publicly Announced Eurocredits for 1974 (millions of U.S. dollars equivalents)

1st Qtr	2nd Qtr	3rd Qtr	4th Qtr	Total* 1974
6,450	12,392	4,575	3,521	28,624

*Credits for which the date of arrangement is not available are excluded from quarterly data but included in the 1974 total. These credits amounted to $1,686 million.
Source: International Bank for Reconstruction and Development.

The volume of new Eurocredits of $19 billion for the first half of 1974 was more than twice the amount for the same period of 1973. Reflecting easy market conditions in the first half, three of the largest Eurocredits were arranged during this period: the $2.5-billion loan for the British Government, the $1.5 billion loan for the French Government, and the $1.2 billion loan for Italy's Mediobanca. Before 1974, the largest Eurocredit was the $1 billion loan for the British Electricity Council, which was arranged in April 1973.

In the first half of 1974, there was a substantial volume of precautionary borrowings by oil-importing countries to accumulate external funds for future financing of oil deficits. In fact, most of the explosive growth of the Eurocredit volume in the first half was accounted for by precautionary borrowings related to the oil crisis. Perhaps as much as three-quarters of the total Eurocredits in this period were due to balance-of-payments considerations. In addition to borrowing themselves and inducing other public-sector entities to do so, governments in a number of oil-importing countries encouraged commercial banks and other enterprises to borrow abroad, especially in the Eurocurrency market. The following table (Table 6.6) shows the high proportion of Eurocredits borrowed by governments and other public entities in the first half.

Table 6.6
Eurocredits by Borrower Category in 1974 (in percent)

	1st Qtr	2nd Qtr	3rd Qtr	4th Qtr	Total 1974
Government & public entities	56	65	30	16	51
Private financial entities	4	3	1	2	3
Public utilities & transport	22	18	30	39	24
Fuels	8	8	14	13	10
Other industry	10	6	25	30	12
Total	100	100	100	100	100

Source: International Bank for Reconstruction and Development.

In early 1974, there was some fear in the market that the expected heavy demand for Eurocurrency funds by oil-importing countries would not only strain the Eurobanking system to a breaking point, but would also lead to a steep rise in Eurocurrency interest rates. (We will discuss the latter point now, while the former is dealt with later.) Before 1974, a sudden heavy demand for Eurocurrencies invariably led to a steep increase in the interest rates of Eurocurrencies, especially Eurodollars. However, this situation has changed fundamentally since the removal by the United States of its various barriers against capital outflows in January 1974, well ahead of the end-1974 deadline which had been announced earlier by the U.S. Government.

The most obvious effect of the lifting of U.S. capital controls has been the narrowing of interest rate differentials between the Eurodollar market and the U.S. money market. Before the lifting of capital controls, arbitrage operations between the two markets had taken only one form, i.e., the inward arbitrage, because while funds could flow into the U.S. market without restriction, they could not flow out freely due to the U.S. controls. As a result, if the demand for Eurodollars increased sharply during an exchange crisis when people wanted to borrow dollars in order to buy, for example, German marks or Swiss francs in the expectation that these currencies would appreciate against the dollar, Eurodollar interest rates shot up sharply, widening the gap between Eurodollar rates and U.S. money rates. For example, in 1971 the premium of the Eurodollar rate climbed to 375 basis points above the U.S. commercial paper rate, as people rushed to borrow Eurodollars to convert them into German marks and Swiss francs. Such a wide differential will not happen again, however, because dollar deposits will be freely transferred to the Eurodollar market from the U.S. market, as soon as Eurodollar rates rise above U.S. money market rates beyond a certain marginal level.[8]

Therefore, the ability of the Eurodollar market to accomodate an increase in the demand for funds by oil-importing countries has grown markedly since early 1974 due to the removal of U.S. controls. As will be elaborated in the following section, this growth in the Euromarket's ability to satisfy increased loan demand has come about without any concomitant decrease in the size of loanable funds in the U.S. market.

EFFECTS OF OIL-MONEY FLOWS ON WORLD CREDIT

In the previous two sections, it was shown that the most important factor behind the growth of the Eurocurrency market in 1974 was the higher price of oil and the subsequent increase in oil-money flows in the form of a higher supply of, and demand for, Eurocurrency funds. To the extent that oil-money flows have contributed to the growth of the Eurocurrency market, they have caused an increase in the availability of credit in the world. Whether the Eurocurrency market has thereby contributed to world inflation depends on whether the increased credit has been used in those areas where there was idle capacity to be utilized.[9] In essence, the size of the Eurocurrency market can increase from two causes: a transfer of bank deposits to the Eurocurrency market from the country of the currency involved; and an increase in the Eurocurrency multiplier, to be explained later.

In the former case, there is an absolute increase in the amount of "primary deposits" being transferred originally out of the country of the Eurocurrency, and the size of the Eurocurrency market will increase by the primary deposits times the Eurocurrency multiplier:

$$\Delta E = k \, \Delta P,$$

where E is the size of the Eurocurrency market, k the Eurocurrency deposit multiplier (to be explained later), and P the amount of primary deposits. Like any other outflow of funds from a Eurocurrency country to the Eurocurrency market, oil-money flows from the United States to the Eurocurrency market do not contract either the U.S. monetary base or the amount of loanable funds in the United States supported by a given U.S. monetary base, while they do contribute to an increase in the lending potential, or the pool of loanable funds, of the Eurocurrency market.

The explanation for the above phenomenon is rather simple. Assume that an OPEC member transfers its dollar deposit held at a bank in the United States to a Eurobank in London, thus creating a Eurodollar deposit. The size of the Eurodollar market increases by the new deposit times the Eurodollar multiplier. In the United States, the OPEC member's original deposit is now transferred to the London Eurobank's deposit account at its bank in the United States. Thus, there is no net change in the total amount of deposits in the U.S. banking system. However, the Eurobanking system has gained new loanable funds from the increased amount in Eurodollar deposits. Thus, the availability of credit in the Eurodollar market has grown accordingly, without any concomitant decrease of credit availability in the U.S. market.

Theoretically, oil-money flows from the United States to the Eurodollar market can affect the U.S. money supply. Suppose

Kuwait transfers a demand deposit of $100 million at Chase-New York to a time deposit at Chase-London, thus increasing the size of the Eurodollar market by $100 million. The U.S. money supply decreases by $100 million in this case, because the $100-million demand deposit of Chase-London at its home office in New York City is not counted as part of the U.S. money supply. In practice, however, Chase-London will usually "spend" its $100 million idle balance on the same day, perhaps within a few hours. Suppose Chase-London lends $100 million to Barclays-London. In this case, $100 million is back into the U.S. money supply, because Barclay-London's demand deposit at a U.S.-based bank is counted as part of the U.S. money supply.

Instead of lending to Barclays-London, Chase-London may lend the money to Chase-New York. In this case, $100 million is again out of the U.S. money supply, because a branch loan to the U.S. head office is not counted as part of the U.S. money supply. Again in practice, however, Chase-New York is not likely to keep an idle balance of $100 million even for one day. Suppose Chase-New York lends the money to the Ford Motor Company in Detroit. This puts $100 million right back into the U.S. money supply, and again there is no actual effect on the U.S. money supply.

The second way that the Eurocurrency market can grow is through an increase in the Eurocurrency multiplier.[10] If oil-money flows contribute to a higher Eurocurrency multiplier, the size of the market will grow even without any increase in the primary deposits. Before discussing the effect of oil-money flows on the Eurocurrency multiplier, we will first define the term "Eurocurrency multiplier." Since most studies on the multiplier deal with Eurodollars, we can substitute the Eurodollar multiplier for the Eurocurrency multiplier without any significant loss to the applicability of the concept. Eurodollars are not money (means of payment) but time deposits; hence analysts generally agree that the Eurodollar multiplier is not a money multiplier. In the U.S. banking system, the money multiplier is an *ex post,* historical ratio of M_1 to the monetary base; the Eurodollar multiplier, on the other hand, is understood either as the Eurodollar deposit multiplier or as the Eurodollar credit multiplier.

The Eurodollar deposit multiplier (k) has generally been defined as an *ex post* ratio of Eurodollar deposits, net of interbank deposits, to a certain "base." Economists at the First National City Bank (FNCB) of New York employ the U.S. monetary base as the "base" of the multiplier. Thus, the FNCB multiplier is expressed:

$$k = \frac{ED}{MB_{us}},$$

where ED is the volume of Eurodollar deposits, as reported by the Bank for International Settlements, net of interbank deposits, and MB_{us} is the U.S. monetary base.[11]

Boyden Lee defines the Eurodollar deposit multiplier as the ratio of foreign-held short-term claims in dollars, including net Eurodollar deposits but excluding Eurobanks' dollar demand deposits with U.S.-based banks, to foreign-held short-term claims in dollars, excluding net Eurodollar deposits but including Eurobanks' dollar demand deposits with U.S.-based banks. The Lee multiplier is defined as:

$$k = \frac{D_f}{B}.$$

$$D_f \equiv D_o + D_c + D_p + D_i + E_n, \text{ and}$$

$$B \equiv D_o + D_c + D_p + D_i + R,$$

where D_o, D_c, D_p, and D_i denote short-term dollar holdings of, respectively, foreign official institutions, foreign commercial banks (excluding dollar demand-deposits of Eurobanks), foreign private non-banks, and international organizations. E_n is net Eurodollar deposits (the same as ED in the FNCB multiplier) and R the dollar demand-deposits of Eurobanks held with U.S.-based banks, described by Lee as "Eurobanks' cash reserves against Eurodollar deposits."[12]

The FNCB multiplier and the Lee multiplier are each a product of efforts by economists to explain the growth of the Eurocurrency market in general and of the Eurodollar market in particular. The Eurodollar deposit multiplier is strikingly similar in concept to the U.S. money multiplier, and it is indicative of the painstaking attempts by economists to apply the analytic techniques developed in U.S. monetary economics to a study of the Eurodollar phenomenon. For example, Milton Friedman argues that Eurodollar institutions are part of a fractional reserve banking system, very much like Chicago banks. According to him, the failure to recognize "the magic of fractional reserve banking" is the chief source of misunderstanding about the Eurodollar market.[13] However, Friedman's propositions confuse what is possible with what has happened in fact. Although in theory credit and deposit creation in the U.S. banking and Eurodollar systems might be similar, in actual practice the forces behind monetary expansion in the two systems differ in many important respects.

Both U.S. banks and Eurobanks incur deposit liabilities which are a multiple of their cash reserves. In this sense, both systems engage in fractional reserve banking, as pointed out by Friedman. But Fred Klopstock notes that this characteristic is common to all financial intermediaries, such as U.S. commercial banks, savings banks, life insurance companies and Eurobanks. All these institutions convert all but a small part of the funds they receive into earning assets. Consequently, their cash reserves are only a fraction of their liabilities. According to Klopstock, however, this fact alone does not explain the striking differences in their credit-creating powers, because keeping fractional cash reserves is not the same as engaging in multiple credit expansion.[14]

With this difference in mind, let us now examine the concept of the Eurodollar credit multiplier. While the Eurodollar deposit multiplier is mainly concerned with the growth of the Eurodollar market in terms of Eurodollar deposits, the Eurodollar credit multiplier is a concept one step ahead of the Eurodollar deposit multiplier as an analytical tool. The Eurodollar credit multiplier is concerned not only with the growth of the Eurodollar market in terms of the market's net assets (the credit side), but also with the Eurodollar market's effect on the volume of world credit. The concept has been developed in the attempt to find answers to these two questions: Can the Eurodollar market create credit internationally and add to the credit base of the world economy? If it can, how and to what extent does it do so?[15]

John Makin approaches the problem with the observation that, in order to extend credit, Eurobanks need to maintain reserves as back-up precautionary balances. He classifies as such reserves the Eurobanks' demand-deposit claims on U.S. commercial banks, exclusive of loans to home offices. Then, he defines the credit multiplier as the ratio of the total Eurodollar assets to the Eurobanks' precautionary reserves, as defined above.[16] His investigation finds a steady increase in the multiplier, a fact which he finds consistent with the effective economies of scale which may be realized in the management of precautionary balances, as the volume of receipts and disbursements of the Eurobanking system rises over time.

In reality, however, few Eurobanks actually keep their demand-deposit claims on U.S. banks for the purpose of using them as back-up reserves or precautionary balances. Eurobanks, as participants in the highly competitive money market system, are very reluctant to keep any of their assets in the form of non-interest-bearing demand deposits, and if they do, they have more likely been "obligated" to do so, as when they maintain compensating balances for dollar bank loans on their lines of credit with U.S. banks. As noted earlier, the idea of demand deposits as a precautionary balance is fast dying out in this age of "liability management" in banking. Any banker who cannot locate interest-earning substitutes for demand deposits with no loss of liquidity is simply not worth his salary. In these days, Eurobanks can take advantage of a host of alternatives to demand deposits as their primary reserves, such as lines of credit with major U.S. banks and investments in highly liquid instruments like Treasury bills, repo loans, and high-quality CDs.

An even more fundamental objection to the Makin credit multiplier arises from the fact that to consider Eurobanks' precautionary balances (either demand deposits or other "back-up" reserves, or both) as the base for Eurocurrency credit expansion is not very useful in evaluating the credit-creating capability of the Eurocurrency market in the absence of regulatory required reserves. Identifying a reserve base is meaningful as an indicator of the maximum potential amount of credit creation only when, as in the U.S. domestic banking system, a compulsory requirement is imposed. In the Eurocurrency banking system, however, a multiplier based on precautionary reserves is devoid of such meaning, because the reserve base of Eurobanks is not under official control and is therefore likely to vary in response to market conditions at the discretion of Eurobanks themselves.[17]

Without a well-defined reserve base, it is difficult to make use of the fractional resrve banking analogy when analyzing the Eurodollar system. The credit or deposit multiplier of a fractional reserve banking system arises because the loans or investments of individual banks do not drain reserves from the system as a whole, and thus lead to an increase in aggregate liabilities by a multiple of the injection of new reserves. For a national banking system, there is no way for cash balances to escape, except into Federal Reserve notes, but for the Eurodollar system there is: back to a deposit in the United States.[18] Therefore, Machlup prefers to do without the concept of the Eurodollar multiplier entirely when analyzing the growth of the Eurodollar market, because the conditions under which this tool can be helpful are absent. Specifically, he notes that "There are no well-specified reserve assets held by Eurobanks, nor do these banks obey any uniform and stable reserve requirements or reserve practices. Thus, we know neither the multiplicand nor the multiplier."[19]

Another, perhaps more relevant way to define a Eurodollar credit multiplier may be to employ the "primary deposits" (defined earlier) as the base. This base is not a reserve base, but at least the multiplier can tell us the degree of credit expansion through the Eurodollar market with a given addition of "new" Eurodollars (or primary deposits). For our study of the effect on the world credit volume of oil-money inflows as primary-deposit transfers to the Eurodollar market, it will be useful to consider a credit multiplier based on the primary deposits. Better still, we may adopt the concept of the marginal credit multiplier (MCM), defined as:

$$MCM \equiv \frac{\Delta EA}{\Delta PD} = \frac{1}{1 - (1-e)\, d},$$

where EA denotes the volume of the Eurodollar credit, measured by the total assets of Eurobanks net of interbank assets, PD the amount of Eurodollar primary deposits, e the proportion of leakages of the primary deposits due to maintenance by Eurobanks of some of their dollar assets in the United States, either as demand deposits or as interest-yielding liquid investments, and d the redeposit ratio of Eurodollar loans, i.e., the proportion of Eurodollar loans which are redeposited in the Eurodollar market by the credit end-users.[20] MCM is not necessarily an *ex post* ratio, but it can be an indicator showing the maximum possible expansion of credit through the Eurodollar market with a given inflow of primary deposits into the Eurodollar market. It should be noted that the higher the leakage ratio (e), the lower the credit multiplier (MCM); and the higher the redeposit ratio (d), the higher the credit multiplier (MCM). As e approaches zero and d approaches 1, MCM approaches the infinite. Conversely, if e approaches 1, MCM approaches 1 regardless of which value d happens to maintain.

Furthermore, we can see that MCM is very similar to the marginal deposit multiplier (MDM) of the Eurodollar market, defined as:

$$MDM \equiv \frac{\Delta ED}{\Delta PD},$$

where ED is net Eurodollar deposit liabilities of Eurobanks. It is not surprising to see that net Eurodollar assets of Eurobanks are almost equal to net Eurodollar deposit liabilities, especially if we assume no currency conversion of Eurodollars by Eurobanks. In our earlier discussion of MCM, we have implicitly assumed that the only way for Eurodollars to disappear from the market is by way of the re-flow of Eurodollar deposits back to the United States, either through Eurobanks' investments in dollar-denominated liquid assets in the United States (leakages), or through no redepositing by Eurodollar credit end-users in the Eurodollar market (absence of redeposits), but not by way of the conversion by Eurobanks of the Eurodollars into other currencies. In reality, the assumption of no currency conversion by Eurobanks does not reduce the practical importance of our multiplier theory, as can be seen by the fact that Eurobanks' Eurodollar assets are practically the same as their Eurodollar deposit liabilities. For example, at the end of December 1974, the Eurodollar deposit liabilities of reporting Eurobanks stood at $155,690 million, while their Eurodollar assets were $155,670 million. Their annual growths were also very similar, leading us to conclude that:

$$\Delta ED \doteqdot \Delta EA,$$

$$MDM \equiv \frac{\Delta ED}{\Delta PD} \doteqdot \frac{\Delta EA}{\Delta PD} \equiv MCM = \frac{1}{1 - (1-e)\,d}$$

Therefore, the relationship of MDM to e and d is about the same as that of MCM to e and d, that is, the higher the leakage ratio (e), the lower the marginal deposit multiplier (MDM); and the higher the redeposit ratio (d), the higher the marginal deposit multiplier (MDM).

How have oil-money flows affected MCM and MDM in the Eurodollar market? We may answer this question by evaluating the effect of oil-money flows on e and d, respectively. OPEC countries tend to deposit their surplus oil revenues with only several scores of major banks, depriving a large number of small and medium-sized Eurobanks of dollar deposits. As oil-money deposits are increasingly concentrated in major banks, many non-OPEC depositors also show a tendency to imitate OPEC countries by transferring their funds from small and medium-sized banks to large banks, perhaps for added security. The banking crisis in 1974 accelerated this process. Consequently, small and medium-sized banks have had to bid aggressively for Eurodollar deposits from large banks in the interbank Eurodollar market, thus increasing interbank transactions in Eurodollar deposits. This may have increased the leakage ratio (e), because the longer the chain of interbank transactions, the greater the proportion of leakages for the original primary deposits.

The leakage ratio (e) may have also increased as Eurobanks, faced with the concentration of Eurodollar deposits in a handful of OPEC countries, tried to keep more of their Eurodollar assets in liquid dollar investments in the United States than in Eurodollar loans. The reluctance of Eurobanks to extend Eurodollar loans may have been caused by two other factors: on the one hand, the increased loan demand for balance-of-payments reasons to finance oil deficits, in comparison to the demand for project loans, which are regarded as the chief domain of Eurobanks; and on the other hand, the increasing pressure on the capital ratio of Eurobanks, as the need for intermediating oil-money flows increases.

Therefore, the overall effect of oil-money flows on the leakage ratio (e) must have been to raise it somewhat. However, many oil-importing countries that have borrowed Eurodollar loans have kept most of their loan proceeds in the Eurodollar market, thus raising the redeposit ratio (d). Hence, it is difficult to measure the net effect of oil-money flows on MCM and MDM; the increase in e tends to lower both MCM and MDM, while the increase in d tends to raise them.

STRUCTURAL CHANGES IN THE EUROCURRENCY MARKET

Oil money had major effects in 1974 on the structure and operation of the Eurocurrency market; in fact, the year will be remembered as a traumatic one for the Euromarket in general and for the Eurocurrency market in particular. At the beginning of 1974, however, the Eurobanking community was euphoric. In January and early February, interest rates had declined substantially. Although there were a number of new borrowers coming to the Eurocurrency market, there was also considerable liquidity to satisfy these borrowers; the Eurocurrency loan market seemed to envision no end to its rapid growth. As noted earlier, in the first half of 1974 the volume of new Eurocredits was growing at twice the rate of the 1973 volume. The average size of individual Eurocredits was increasing also. In the first months of 1974, the average loan approached $90 million, compared to $60 million in 1973 and $40 million in 1972.

During the summer, however, problems began to develop. There were several reasons apart from oil money why the Eurobanking system witnessed in 1974 one of the severest banking crises in the post-war period. From the spring of 1974, interest rates rose sharply in most countries, as governments pursued tight monetary policies to combat inflation. Credits were becoming scarce; banks were hard pressd to acquire new funds. At the same time, Eurobanks recognized the size of the financing burden of oil-importing countries. The balance-of-payments implications of the four-fold increase in oil prices were unsettling for many oil-importing countries. Eurobanks imposed loan limits on most countries, leading to further rigidity in the international financial market.

On the other hand, OPEC countries showed a preference for depositing their newly-acquired resources only with the score or so of the largest banks, thereby depriving many medium-sized and small Eurobanks of deposits. These smaller banks had to pay premiums over the normal interbank Eurocurrency rates to acquire funds, when funds were available to them. This created a multi-tiered rate system in the Eurocurrency market. OPEC countries also preferred to deposit funds for very short-term maturities of perhaps a few days, or up to a few months at most. This, in turn, put pressure on Eurobanks to intermediate very short deposits into medium-term loans for their customers. Many banks simply gave up the role of intermediation, thus further tightening the market.

As deposits were being concentrated in the largest banks, some medium-sized and small banks faced increasing difficulties in finding funds even for rolling over their previous floating-rate Eurocredits; some had to sell these commitments to their parent banks or others. These banks also were losing money on the loans, because the spreads were tied to the "prime" interbank Eurocurrency rate, which they could not get any longer. Furthermore, the interest rate was reset only every six months, and in the interim the cost of borrowing the short-term deposits rose dramatically. At the same time, some banks suffered losses in their foreign exchange operations, mainly because their foreign-exchange traders were not familiar with the new floating exchange-rate system and took imprudent risks. Heavy losses and a few bank failures related to these and other problems created a crisis of confidence in the Eurobanking community. The following table (Table 6.7) shows some of the major foreign exchange losses suffered by financial institutions.

Table 6.7
Foreign Exchange Losses for Financial Year 1974 (in millions of U.S. dollars)

Name of Bank	Amount of Losses	Cause
Franklin National Bank	370	Foreign exchange & loan losses
Westdeutsche Landesbank Girozentrale	114	Foreign exchange dealings
Union Banks of Switzerland	56	Foreign exchange dealings
Lloyds Bank	77	Foreign exchange losses at its Lugano Branch
Banque de Bruxelles	100	Foreign exchange dealings
Bankhaus Herstatt	?	Foreign exchange dealings

Source: *The Banker,* June 1975.

Tight credit conditions also adversely affected bank clients, especially real estate firms. Consequently, loan losses were increasingly experienced by Eurobanks, and bank-loan portfolios had to be reevaluated by bank managements. Many borrowers, especially those from developing countries, could no longer obtain credits from Eurobanks because of the banks' credit-upgrading operations and their shift into alternative higher-credit investments.

In the ensuing crisis, consortium banks in the Eurobanking system were hard hit. Consortium banks, which are set up by international groups of commercial and investment banks, usually specialize in Eurocurrency loan business. According to some observers, the emergence of these multinational consortium banks, which first appeared in 1964, is the most important development in banking since the end of World War II. There is a great diversity among consortium banks both in terms of their style of operations and the kinds of services offered to their clients. Due to the heterogeneity of these banks, some regard them as a mixture in varying proportions of both commercial and investment banks. Today practically all international banks have participated in establishing one or two consortium banks. By and large, shareholders of most consortium banks are themselves banks, except in a few cases.[21]

Until the Eurobanking crisis of 1974, consortium banks had achieved an impressive growth record by actively engaging in Eurocredit business. However, the 1974 banking crisis caused serious disruptions to consortium banks. In an atmosphere of fundamental doubt, some depositors in the Eurocurrency market were concerned that the parent banks might not support their consortium bank, in which they usually held only a minority share, sometimes as low as ten percent. This fear was heightened by the fact that the loan portfolios of consortium banks had grown rapidly in previous years, without concomitant increases in their deposit or capital base. These loans were largely supported by active borrowings in the interbank Eurocurrency market, since consortium banks, by their structure, have neither depositing-gathering branches nor wide commercial contacts, and thus are forced to fund themselves almost exclusively through the wholesale Eurocurrency interbank market. Furthermore, there was some uncertainty, for a time, whether the central bank in the country of incorporation would feel obliged to provide support for such essentially "offshore" operations as consortium banks.

The crisis of confidence in the Eurobanking community was a matter of concern to many central banks. In a series of meetings, central bankers of major Western countries agreed on cooperative measures to reaffirm their commitment to stand behind their banks in liquidity difficulties. Furthermore, at the urging of the Bank of England, the parent banks of London-based consortium banks explicitly pledged to provide support to their consortium banks, if their support was required. Though these pledges of support were couched in fairly general terms, and each bank was free to choose its own phrasing, the positive response of the parent banks was regarded as a firm moral commitment to support their respective consortium banks. The formal reaffirmations of shareholder responsibility towards consortium banks tended to change the character of these banks from a limited-liability company to a partnership, if not *de jure,* then *de facto.*

Due to the Eurobanking crisis, the average maturity of Eurocredits in 1974 declined noticeably. While almost one-fifth of the total Eurocredit volume in 1973 had a maturity of more than ten years, less than one-tenth belonged to that category in 1974. The unstable market condition in the second half, following a period of comparatively easy borrowings in the first half, is also well illustrated by the wide difference in premiums in 1974 compared to 1973. For example, prime borrowers from developed countries paid as low as ⅜ percent for the premium (or spread) over the London interbank offer rate, called the LIBO, in the early months of 1974. In the second half, however, many Eurobanks simply refused to participate in any Eurocredit carrying less than 1 percent for the spread.

Not only did the spreads for the same class of borrowers experience wide fluctuations in 1974, but the mode of fixing the

Table 6.8
Publicly Announced Eurocredits by Maturity (in percent)

Maturity	1973	1974
1-6 years	12%	15%
7-10 years	56%	69%
Over 10 years	18%	8%
Unknown	14%	8%
Total	100%	100%

Table 6.9
Typical Eurocredit Spreads over LIBO (in percent)

	1973		1974	
	Spreads	Difference	Spreads	Difference
Prime borrowers from developed countries	1/4-5/8	3/8	3/8-1	5/8
Public borrowers from socialist countries	1/2-3/4	1/4	5/8-1 1/2	7/8
Prime LDC borrowers	3/4-1	1/4	7/8-1 1/2	5/8
Non-prime LDC borrowers	1 1/4-1 3/4	1/2	1-2	1

spread itself underwent an important change. Early in the year, both lenders and borrowers understood perfectly well the concept of the LIBO, over which the spread of a Eurocredit would be set. In the second half of the year, however, the London interbank Euromarket developed a multi-tiered rate system due to the banking crisis. Thus, it became important in a Eurocredit contract to clearly note the reference banks whose LIBO would be used as *the* LIBO for the specific Eurocredit.

Furthermore, instead of a single spread for the entire maturity of a Eurocredit, the multi-spread method was introduced to protect lending Eurobanks from future cost escalation. For example, a Eurocredit with a seven-year maturity may carry a spread of ⅞ percent for the first three years, a spread of 1 percent for the following two years, and a spread of 1½ percent for the final two years. However, some Eurobankers expressed doubt over the efficacy of such "step-up" spreads (i.e., successively *larger* spreads) in adequately protecting the Eurobanks from a profit squeeze arising from higher future costs, because most borrowers could repay the loan any time without penalty. As long as there is a prepayment-without-penalty clause, the utility of step-up spreads is greatly reduced. Several years from now, for example, certain borrowers may simply refund their existing Eurocredit containing step-up spreads with a new Eurocredit, rather than paying the higher step-up spread taking effect from that time. In order to remedy this problem, some bankers have suggested "step-down" spreads; in this case, a Eurocredit would have successively *lower* spreads. As a result, borrowers would be discouraged from refunding a Eurocredit in the future merely to take advantage of any spread differences.

Even during the gloomiest days of 1974, however, few really doubted the validity and usefulness of the Eurocredit concept. The fundamental *raison d'etre* of Eurocredits is that the concept of floating-rate roll-over loans allows the Eurobanking community to increase substantially its intermediation role in financial markets. Because of recent inflationary experiences as well as the accelerated velocity of international funds, Eurocurrency deposits in general are available in short maturities—probably at the expense of medium or long-term deposits. On the other hand, the demand for Eurofunds has been increasingly for longer maturities. The only way for Eurobanks to successfully meet these conflicting needs of the suppliers and the demanders of Eurofunds is to utilize the Eurocredit concept of floating-rate roll-over loans.

Furthermore, since individual loan requirements in the international market overwhelm the lending power of any particular Eurobank, the principle of bank syndication in Eurocurrency financing is indispensable. The institution of the consortium bank is particularly well suited to meet this challenge in the Euromarket, though during the height of the banking crisis in 1974, some people questioned the very concept of consortium banks. Both the synergetic and risk-spreading functions inherent in consortium banks can complement the international activities of their parent banks and should, moreover, play a specially useful role in strengthening the Euromarket.[22]

In addition, the 1974 crisis gave Eurobanks a new direction in their banking philosophy.[23] Obviously, the idea of growth for its own sake is no longer valid, and many banks now pay close attention to profitability. Specialization, geographical or functional, is another trend apparent in Eurobanking. Those Eurobanks that possess special expertise in a certain field are expected to play a growing role in loan appraisal as well as loan syndication. New and closer cooperation with other centers of international finance, particularly the Middle East, is also becoming important. Furthermore, the increased role of oil money in the Eurocurrency market has enhanced the position of OPEC banks in this market. Several banks from Kuwait and Saudi Arabia as well as Venzuela now play a prominent role in the syndication of Eurocredits.

ROLE OF THE EUROCURRENCY MARKET IN RECYCLING OIL MONEY

We have seen that oil-money flows contributed significantly to the growth of the Eurocurrency market in 1974. In an earlier section, we have also noted that a substantial number of Euroloans have been raised by oil-importing countries to finance oil deficits. The existence of the Eurocurrency market has thus aided the recycling of oil money. To the extent that the Eurocurrency market has fulfilled the recycling task, it has helped to reduce the painful adjustment costs of many oil-importing countries in terms of deflationary economic policies, external trade barriers, competitive currency depreciations, and other "beggar-thy-neighbor" policies.

While recognizing the beneficial effects of oil-money recycling by the Eurocurrency market, we should also note its limitations as a recycling channel. First, the Eurocurrency market, being a free private financial market, is a difficult place to borrow money for [illegible] eveloping countries that may carry a low credit [illegible] nevertheless in most serious need of oil money. [illegible] nal aspect remains an economic, as well as poli- [illegible] ge of Euromarket recycling. The credit-worthiness [illegible] is not necessarily correlated to the welfare effects [illegible] terms of economic efficiency. Also, there is a defin- [illegible] e extent to which the Eurocurrency market can play a [illegible] ole due to the inadequacy of the capital base of most Eurobanks. The task of oil-money recycling is simply too big for the Eurocurrency market to perform singlehandedly or, for that matter, too big for any free-market mechanism.[24] The basic problem with the recycling of oil money by the Eurocurrency market is that the loans involved are not self-liquidating. Unlike a project loan or a trade credit, a bank loan to finance oil imports belongs to the consumption loan category; the loan enables an oil-importing country to buy oil and consume it. For poor developing countries with low foreign-exchange reserves, such consumption loans may have very low repayment priority.

Therefore, the world still needs some sort of extra-market recycling mechanism to complement free-market recycling systems, including the Eurocurrency market. In this sense, the establishment of the IMF oil facility, the OECD safety-net fund, and the World Bank's Third Window should all be regarded as moves in the right direction. In general, it may be perfectly normal for a payments deficit which is financed in its early stages from private market sources, including the Eurocurrency market, to be subsequently refinanced through a compensatory official borrowing mechanism. History has shown that without effective official measures in the end to cope with the deficit problem, the private market will reverse its earlier constructive and highly effective equilibrating role and become a severe disequilibrating force in the later stages of large and sustained deficit.[25]

EFFECTS OF OIL MONEY ON THE EUROBOND MARKET

The Eurobond market was in a depressed state in 1974 due to high short-term interest rates and a reverse yield curve which lasted most of the year. However, the market picked up considerable steam after the fall of 1974, when short-term rates came down as a result of reflationary monetary policy in most industrial countries. The volume of new Eurobond issues increased sharply from the beginning of 1975, as indicated in Table 6.10.

Table 6.10
New Eurobond Issues, 1972-75 (millions of U.S. dollars)

1972	1973	1974					1975		
		1st Q	2nd Q	3rd Q	4th Q	Total	1st Q	2nd Q	1st Half
6,928	4,615	928	743	1,167	1,639	4,477	2,674	2,801	5,475

Oil-money flows have influenced the Eurobond market in two ways. First, financial institutions in oil-producing Arab countries have emerged as a major force in the Eurobond underwriting business. Since 1974, most Eurobond issues have had active Arab bank participation, frequently as lead managers.[26] Second, the Eurobanking community has been eager to attract Arab investors to the Eurobond market. Many Arab investors have shown a preference to purchase long-term bonds denominated in their own currencies, rather than in "weak" currencies such as the U.S. dollar. Subsequently, there has been a growing use of Arab currencies in the market.[27] On the other hand, however, borrowers are reluctant to incur long-term debts denominated in "strong" Arab currencies, which are likely to appreciate in the future. As a compromise, the Eurobanking community has created an artificial currency unit composed of various Arab currencies.

The new composite currency unit, called the Arcru (the Arab Currency-Related Unit), was first introduced in November 1974. The Arcru is composed of these twelve Arab currencies:

Algerian dinar
Bahrain dinar
Egyptian pound
Iraqi dinar
Kuwaiti dinar
Lebanese pound
Libyan dinar
Oman riyal
Qatar riyal
Saudi Arabian riyal
Syrian pound
United Arab Emirates dirham

However, the number of component currencies used in *actually* estimating the value of the Arcru is only eight, because the two strongest and two weakest component currencies since the base date (June 28, 1974) are eliminated before making the estimate. On the base date of June 28, 1974—chosen so as to coincide with the base date of the new SDR—the Arcru had a value of SDR 0.828945, or US $1. From that date on, the value of the Arcru in terms of the U.S. dollar has fluctuated to reflect the average of the subsequent movements in the dollar exchange rates of the middle eight currencies. Unlike other currency-basket units such as the SDR or the Eurco, however, the middle eight currencies actually used in calculating the value of the Arcru have equal weights, each contributing 12.5% in value to one Arcru.

The composition of the eight middle currencies will change gradually, as the relative strengths of the twelve component currencies vis-a-vis the U.S. dollar change over time. In practice, the Arcru is little different from the U.S. dollar; all the transactions (loan subscriptions as well as interest and principal payments) are effected in U.S. dollars for an Arcru loan. The only difference between the Arcru and the U.S. dollar on a given date is the divergence (or gap) of the average value of Arab currencies vis-a-vis the U.S. dollar from the base date ratio of June 28, 1974, when one Arcru equaled one U.S. dollar. This difference is analogous to that between the current dollar and the constant dollar. While a constant dollar is taken to mean an actual dollar adjusted for price changes since 1967, the Arcru is the actual dollar adjusted for exchange rate changes in Arab currencies since June 28, 1974.

The dollar value of the Arcru will be calculated daily, as is the value of the SDR by the IMF in major currencies, and the value of the Eurco by the Luxembourg Stock Exchange in EEC and U.S. currencies. For this purpose, Hambros Bank, which originally introduced the Arcru, has made provision, in cooperation with several Arab-related banks, for the regular receipt of the necessary exchange-rate information. The dollar value of one Arcru is calculated in the following way. First, an estimate is made of the percentage changes from the base date in the exchange rates of the 12 component currencies vis-a-vis the U.S. dollar. Next, the two strongest and two weakest currencies are excluded from the twelve. Finally, a calculation is made of the simple average change in the exchange rates of the remaining eight currencies vis-a-vis the dollar.

The Arcru has two advantages over other composite currency units such as the SDR and the Eurco. The inclusion of only the middle eight currencies out of the twelve eliminates the effects of extreme aberrations in the exchange rate of any one currency, whether temporary or persistent. This feature might be attractive to borrowers in particular. Additionally, since the Arcru is based on Arab currencies, it might be attractive to Arab investors who would like to make loans in their own currencies to avoid foreign exchange risks. Thus, the Arcru appears to provide a happy middle ground between non-Arab borrowers and Arab investors.

There are some disadvantages, too. Of the twelve Arab currencies, some are pegged to the U.S. dollar, while others are pegged to the SDR or similar composite currency units.[28] Since the value of the Arcru depends on the actual exchange rates between Arab currencies and the U.S. dollar, the influence of many Arab currencies not pegged to the U.S. dollar may increase the volatility of the Arcru value more than if the group of currencies were all pegged to the U.S. dollar. In addition, there is no guarantee that several Arab creditor countries will not engage in manipulating their exchange rates against the dollar to influence the value of the Arcru at the time of loan repayment.

NOTES

[1]For a detailed view on these points, see Carl H. Stem, "The Eurocurrency System: Problems and Prospects," (Paper presented to a Eurocurrency Conference at the American Institute for Public Policy Research, Washington, D.C., October 17-18, 1974.)

[2]*The Oil Situation and the Eurocurrency Market,* Bank for International Settlements, (Basle, Switzerland, February 1, 1974.)

[3]*Forty-fifth Annual Report,* Bank for International Settlements, (Basle, Switzerland, June 1975), p. 130.

[4]*Ibid.*

[5]*Ibid.*, p. 140.

[6]*Ibid.*, p. 139.

[7]Fritz Machlup, "Five Errors about the Eurodollar System," *Euromoney,* July 1972.

[8]For a more detailed explanation of this point, see John Atkin, "Eurodollar Interest Rates," *The Banker,* February 1973; and J. M. Atkin, "U.S. and Euromarket Interest Rates—A Measure of the Euromarket's Health," *Euromoney,* December 1974.

[9]For a rather definitive view that the Eurocurrency market *has* contributed to world inflation, see Jacques Rueff, "How the Euromarkets Contribute to World Inflation," *Euromoney,* April 1975. For an opposite view, see Jacques Riboud, "Why Rueff Is Wrong Over the Euromarkets," *Euromoney,* September 1975.

[10]For a skeptical view on this point, see Helmut W. Mayer, "Some Analytical Aspects of the Intermediation of Oil Surpluses by the Eurocurrency Market," *Banca Nazionale del Lavoro Quarterly Review,* September 1974.

[11]*Monthly Economic Letter,* First National City Bank (New York, July 1974.)

[12]Boyden E. Lee, "The Eurodollar Multiplier," *The Journal of Finance,* September 1973, pp. 867-69.

[13]Milton Friedman, "The Euro-Dollar Market: Some First Principles," *The Morgan Guaranty Survey,* October 1969.

[14]Fred Klopstock, "Money Creation in the Eurodollar Market—A Note on Professor Friedman's Views," Federal Reserve Bank of New York *Monthly Review,* January 1970.

[15]E. Wayne Clendenning, "Eurodollars and Credit Creation," *International Currency Review,* March/April 1971, p. 12.

[16]John H. Makin, "Identifying a Reserve Base for the Eurodollar System," *The Journal of Finance,* June 1973.

[17]A similar view is expressed in John Hewson and Eisuke Sakakibara, "The Euro-Dollar Deposit Multiplier: A Portfolio Approach," *IMF Staff Papers,* July 1974.

[18]Keith Josephson, "The Eurocurrency Market," *The Journal of Commerce,* September 30, 1974.

[19]Fritz Machlup, "The Eurodollar System and its control," in *International Monetary Problems,* Papers and Proceedings of a Conference at

the American Enterprise Institute for Public Policy Research, Washington, D.C., September 1967, pp. 9-11.

[20]*Eurocurrency Markets,* International Monetary Fund, (Washington, D.C., September 20, 1974), p. 14.

[21]For a detailed explanation of consortium banks and their role, see Yoon S. Park, "Structure and Function of the Eurocredit Market," *Euromoney,* April 1974.

[22]For more discussions on this topic, see Yoon S. Park, *Ibid.;* and Alberto Weissmuller, "London Consortium Banks,"*Journal of the Institute of Bankers,* August 1974.

[23]Yoon S. Park, "A New Philosophy for Euromarket Banks," *Euromoney,* June 1975.

[24]For a similar view, see Polly Reynolds Allen, "Can the Eurocurrency Market Finance the Oil Deficits?" *The Banker,* November 1974.

[25]Fred Hirsch, "The Oil-Financing Conundrum," *The World Today,* March 1975, p. 110.

[26]Hassan Abbas Zaki, "Arab Financial Institutions and a New Financial Market," *Euromoney,* July 1975.

[27]For details, see "The Growing Use of Arab Currencies," *The Banker,* March 1975, pp. 287-291.

[28]Michael Melville, "The Arcru: a Progress Report," *Euromoney,* July 1975.

Chapter 7
Effects on International Liquidity and Inflation

Apart from its deflationary effect on the world economy, the rise in oil prices has had a serious impact on the world balances-of-payments condition and international reserves. The average price of OPEC's crude-oil exports quadrupled in the 1973-74 period, and the value of OPEC oil revenues is estimated to have risen from about $32 billion in 1973 to $115 billion in 1974, according to the BIS Annual Report, which estimated the oil revenues on the accrual basis rather than on the cash basis. This increase in OPEC oil revenues transformed the world's trade balances in 1974, as shown in the following table. Table 7.1 (p. 122) shows that, predictably, the steep oil-price increases in 1973 and 1974 exacerbated the world's trade imbalances in 1974.

In 1973, the OPEC member countries as a group already enjoyed a trade surplus of $23 billion, but this surplus rose to $103 billion in 1974. On the other hand, the trade deficits of industrial countries and non-oil LDCs further deteriorated in 1974, mostly due to increased payments for oil imports. The larger burden of the OPEC surplus fell upon the industrial countries, whose economies are relatively more energy-intensive, and thus require more oil imports, than non-oil developing countries. Among the inustrial countries, however, trade performances differed markedly, reflecting the fact that the burden of added oil costs fell unequally upon these countries, and their non-oil trade balances also behaved differently. These developments can be seen in Table 7.2 (p. 122) for the Group of Ten (G-10) countries.

Table 7.1
World Trade Balances in 1973 and 1974 (millions of U.S. dollars)

	Industrial* Countries	OPEC** Countries	Non-oil LDCs
1973:			
Exports (fob)	391,884	44,900	87,816
Imports (cif)	401,035	21,900	111,165
Trade balance	-9,151	+23,000	-23,349
1974:			
Exports (fob)	521,380	138,900	121,840
Imports (cif)	567,600	35,800	177,490
Trade balance	-46,220	+103,100	-55,650

*Seventeen countries consisting of the Group of Ten member countries, plus Austria, Australia, Denmark, New Zealand, Norway, South Africa, and Switzerland.

**Including a few other small oil-exporting LDCs such as Brunei and Trinidad and Tabago.

Source: *International Financial Statistics,* IMF, Washington, D.C., August 1975.

Table 7.2
Trade Balances and Factors of Change for G-10 Countries (billions of U.S. dollars)

Countries	Trade balances*		Changes in trade balances 1973–74	Factors of change			
				Changes in net oil expenditure	Changes in non-oil trade		
	1973	1974			total value	owing to volume	owing to terms of trade
	in milliards of US dollars						
Belgium	0.5	− 1.5	− 2.0	− 1.7	− 0.3	− 0.2	− 0.1
Canada	1.9	0.5	− 1.4	+ 0.4	− 1.8	− 2.1	+ 0.3
France	− 1.4	− 6.9	− 5.5	− 6.4	+ 0.9	+ 1.8	− 0.9
Germany	12.9	19.7	+ 6.8	− 6.3	+13.1	+ 8.4	+ 4.7
Italy	− 5.6	−10.7	− 5.1	− 5.6	+ 0.5	+ 3.2	− 2.7
Japan	− 1.4	− 6.8	− 5.4	−14.2	+ 8.8	+ 5.6	+ 3.2
Netherlands	0.3	− 1.6	− 1.9	− 0.9	− 1.0	+ 1.3	− 2.3
Sweden	1.5	0.1	− 1.4	− 1.4	0.0	− 0.8	+ 0.8
Switzerland	− 2.1	− 2.5	− 0.4	− 0.5	+ 0.1	+ 0.5	− 1.1
United Kingdom	− 8.3	−15.5	− 7.2	− 5.7	− 1.5	+ 1.1	− 2.6
United States	1.3	− 2.3	− 3.6	−16.3	+12.7	+ 7.0	+ 5.7

* Imports c.i.f.

Source: *Forty-Fifth Annual Report,* Bank for International Settlements, Basle, June 1975, p. 22.

Tables 7.1 and 7.2 make it clear that changes in the world's trade balances from 1973 to 1974 were strongly influenced by the sudden increases in the price of oil in late 1973 and early 1974. The oil-price rise produced a radical transformation of the global pattern of trade balances. While the OPEC member countries, whose combined trade balance was already in surplus in 1973, were able to raise further their collective trade surplus to $103 billion in 1974, seventeen industrial countries suffered a combined trade deficit of $46 billion in 1974, compared to $9 billion in 1973. Non-oil developing countries suffered almost as much, their collective trade deficit worsening to $56 billion in 1974 from $23 billion in 1973.

The year 1974 was also the fifth consecutive year of rapid global reserve expansion. International reserves consist of gold holdings by monetary authorities, Speical Drawings Rights held by IMF member countries, Reserve Positions in the IMF arising from countries' gold subscriptions to the IMF and from the IMF's use of members' currencies to finance the drawings of others, and foreign exchange. Foreign exchange is defined as holdings by monetary authorities (central banks, currency boards, exchange stabilization funds, and treasuries to the extent that they perform similar functions) of claims on foreigners in the form of bank deposits, treasury bills, short and long-term government securities, and other claims usable in the event of a balance of payments deficit. The latter include nonmarketable claims arising form inter-central bank and intergovernmental arrangements, without regard to whether the claim is denominated in the currency of the debtor or the creditor.[1] Total international reserves rose by $35.7 billion, or by 18 percent, in 1974. Reflecting the asymmetry in trade balances, however, most of the reserve expansion accrued to OPEC countries, as shown in Table 7.3 (p. 124).

A major reason why total international reserves expanded rapidly in 1974 was that the process of international reserve creation took place when the reserves transferred to the oil-exporting countries were replenished by oil-importing countries through

Table 7.3
Changes in Global Reserves (millions of U.S. dollars)

		Foreign Exchange	Gold, SDRs, & IMF positions	Total Reserves
Seventeen industrial countries	1973	4,471	4,786	9,257
	1974	285	2,363	2,648
OPEC countries*	1973	3,325	312	3,637
	1974	30,729	1,953	32,682
Non-oil LDCs	1973	10,399	1,030	11,249
	1974	494	-208	286
All countries	1973	18,195	6,128	24,323
	1974	31,508	4,108	35,616

*Including a few non-OPEC oil-exporting countries.

Source: *International Financial Statistics,* August 1975.

borrowing in the United States or the Eurocurrency market. In this way, both the United States and the Eurocurrency market have enlarged their role as the principal reserve suppliers to the world as a result of the oil-price increase. Especially, the removal in early 1974 of U.S. capital controls encouraged dollar outflows from the United States to oil-importing countries as well as to the Eurocurrency market. On January 29, 1974, the U.S. government abolished three controls: the Interest Equalization Tax, the Foreign Direct Investment Program, and the Voluntary Foreign Credit Restraint Program.[2] This action by the U.S. Government freed U.S. investors and financial institutions to lend abroad in search of substantially higher yields. As shown in Table 7.4 (p. 125), U.S. banks alone increased their foreign lendings by almost $20 billion in 1974.

The rate of the world's reserve increase of 18 percent in 1974 was comparable to that recorded in 1972 (19 percent) and substantially higher than that in 1973 (4 percent). During the first quarter of 1975, however, the rate of reserve growth slowed down to an annual rate of 9 percent. Additions to foreign exchange holdings accounted for nine-tenths of the total increase in

Table 7.4
Claims on Foreigners Reported by Banks in U.S.A. (Amounts outstanding; in millions of dollars)

End of period	Short-term claims	Long-term claims	Total
1971	13,272	3,667	16,939
1972	15,676	5,063	20,739
1973	20,723	5,996	26,719
1974	38,889	7,155	46,044
1975 (May)	45,028	7,852	52,880

Source: *Federal Reserve Bulletin,* July 1975.

reserves in 1974 and also dominated the reserve increase in the first quarter of 1975. Almost the entire reserve increase recorded in 1974 accrued to OPEC countries. On the other hand, the reserves of seventeen industrial countries as a group showed an increase of only $2.6 billion, while those of non-oil LDCs hardly increased at all. As a result, the ratio of reserves to the volume of annual imports deteriorated sharply for non-OPEC countries in 1974, while the exact opposite was the case for OPEC countries as a group (see Table 7.5).

Table 7.5
Ratio of International Reserves to Annual Imports

	1973	1974
Seventeen industrial countries	31%	22%
OPEC countries*	68%	133%
Non-oil LDCs	39%	25%

*Including a few small non-OPEC oil-exporting LDCs.
Source: *International Financial Statistics,* August 1975.

It has been asserted that the build-up of international reserves by OPEC countries should not be considered in the same light as the accumulation of official monetary reserves by other surplus countries. The increase in OPEC reserves is seen as a transfer of national wealth from the form of oil in the ground into the form of international reserves or financial claims on the rest of the world. Whatever their nominal form, so goes this argument, most of the so-called reserves of OPEC member countries are more analogous to the long-term investments abroad held by U.S. corporations, or the portfolio of common stocks held by the British Government for many years.[3] Yet conceptually, the international reserves being accumulated by OPEC countries are no different from those accruing to other chronic surplus countries. The monetary and financial effect of such a massive transfer of reserves to a handful of countries from the rest of the world is always destabilizing, regardless of whether the surplus countries are oil-exporting countries or industrial countries.

The skewed distribution of international reserves between OPEC and non-OPEC countries raises the question of the world's reserve adequacy. At first glance, the fact that international reserves increased sharply in 1974 may give the impression that the global adequacy of reserves has not been impaired by the oil crisis. Furthermore, the widespread adoption of a floating exchange-rate regime, as well as the enhancement of both the ability and the willingness of governments to borrow in private markets, especially the Eurocurrency market, may appear to support the view that global reserve needs might have been substantially reduced in the past couple of years.

However, the evidence supporting such a view is more apparent than real. Although the nominal value of global reserves increased by 18 percent during 1974, real reserves (i.e., reserves deflated by an appropriate price index) declined, as unit values of traded goods rose by 35-40 percent.[4] The existence of many fixed exchange-rate arrangements, such as the European Currency Snake, and the pegging of most LDCs' currencies to the currencies of their dominant trading partners, may also attenuate the power of a floating-rate regime to lessen the equilibrium size of global reserves. The majority of developing countries have continued to maintain a fixed link to a major currency because of the advantage of administrative simplicity; it avoids the process of continuous decision making that would be necessary under a floating exchange-rate regime. In addition, it ensures that trade denominated in the major currency is conducted at a stable exchange rate, and that stable cross rates obtain between countries which peg their currencies to the same major currency.

Since the Smithsonian agreement of December 1971 ceased to be effective in March 1973, countries have adopted a variety of exchange regimes, depending on their particular circumstances and needs. The exchange rate practices as of 1975 can be classified roughly under the following five categories: (i) independent floating; (ii) pegging to a single currency; (iii) pegging to a composite (or basket) of currencies, including the Special Drawing Right; (iv) pegging to a single currency, with frequent changes in the peg according to a predetermined formula, e.g., based on relative inflation rates; and (v) joint floating under mutual intervention arrangements. All the industrialized countries fall under either category (i) or category (v). Non-industrial countries for the most part have elected to peg their exchange rates, and thus come under categories (ii), (iii), and (iv). The majority of developing countries have maintained a fixed link with the currency in which they traditionally intervene in the exchange market.[5]

Access to private markets does not help many countries with a deteriorating payments condition. International bankers have become increasingly reluctant to extend loans covering balance-of-payments deficits, and many countries have nearly exhausted the country or geographical loan limits imposed by most lending banks. When even a major industrial country such as Italy could not borrow in private international markets in 1974 due to its deteriorated b.o.p. condition, it is quite evident that access to capital markets is severely limited for most of the non-oil LDCs. Finally, the more pronounced unevenness of reserve distribution has caused an increase in the equilibrium size of global reserves. The conclusion to be drawn from all the above factors appears to be that the adequacy of global reserves has deteriorated, rather than improved, since the beginning of the oil crisis.

This conclusion should not necessarily lead us to ask for an increase in the supply of international reserves to match whatever the increased demand for them may turn out to be. Already the steep rise in oil prices has greatly exacerbated worldwide inflation, and further creation of international reserves, whether in the form of SDRs, IMF credits, or other facilities, can only worsen the situation. It has been convincingly demonstrated that there exists a systematic relationship between changes in international reserves and the rate of worldwide inflation.[6]

There has been some confusion as to whether the steep increases in oil prices have accelerated or decelerated world inflation. On the one hand, the oil-price increases add to cost-push inflation, petroleum being directly or indirectly an important component of both wholesale and retail price indices. Thus, the prices of both oil products and their substitutes have increased. Another inflationary push comes from the fact that,due to a deteriorating external payments condition, the currencies of many oil-importing countries may depreciate against the currencies of OPEC and even against those of other oil-importing countries, thereby raising the prices of their imports or import substitutes. In contrast, prices in the countries with appreciating currencies are likely to be sticky downward, thus in effect producing a ratchet effect in world prices. In other words, prices in countries with depreciating currencies go up, while prices in countries with appreciating currencies do not fall as much or even stay the same. An inflationary push also comes from labor's attempt to catch up with price inflation by demanding higher wages.

While the oil-price increases can thus contribute to world inflation, they also have a deflationary effect analogous to the excise tax. This question has been discussed in detail in Chapter 5. Suffice it to note here that, since the real income of most people in oil-importing countries *must* fall due to the oil-price hike, the general price level in these countries will likely be permitted to rise enough to absorb the increase in oil prices, given the difficulty of reducing real wages by lowering nominal wages. In other words, since both nominal wages and nominal prices have been increasingly sticky on the downside, price increases will have to play a proportionately larger role in the income adjustment process, and this is clearly inflationary.

In sum, there can be no doubt about the adverse effect of oil-price hikes on world inflation. By the fall of 1973, even before the steep increase in oil prices was put into effect, the rate of inflation in the OECD countries, as measured by the consumer price index, was already running high at an annual rate of over 8 percent, compared to 4.6 percent for the year 1972. Several factors contributed to this acceleration of inflation in 1973, the most important being the general and very rapid expansion in the world's aggregate demand in 1972 and 1973, the implementation of expansionary monetary and fiscal policies during those years, the emergence of supply bottlenecks during the 1973 economic boom, and increased food prices due to worldwide drought and other natural calamities. Then, in late 1973, the prices of crude oil quadrupled in just a few months. This quantum jump in the price of a commodity as pervasively used as petroleum quickly led to substantial increases in the production cost of many goods and services.

Whereas increases in food prices had been the single most important factor in pushing up inflation rates in 1973, oil prices played the major role in the doubling of inflation rates between late 1973 and the fall of 1974. For the United States, the overall inflationary impact from oil-price increases has been estimated at 2 percent for the GNP deflator and a little more than 3 percent for the cost-of-living index.[7] Worldwide, it has been claimed by an OPEC official that the oil-price hikes of 1973-74 may be responsible for only 2 to 3 percent of the global inflation rate of over 10 percent.[8] According to estimates made by the OECD Secretariat, one-fifth of the overall rise in consumer prices that occurred in the OECD area during the six months from September 1973 to March 1974 was attributable to the oil-price increases. As the higher costs of oil worked their way more fully through various sectors of the world economy in the subsequent months of 1974, prices of finished goods were more widely affected, public utility charges rose, and prices of oil substitutes also increased. Consequently, the overall impact of the oil-price hikes on worldwide inflation may be even larger than the one-fifth share estimated for the initial period following oil-price increases.[9]

NOTES

[1]See "Introduction" in any issue of *International Financial Statistics,* a monthly published by International Monetary Fund, Washington, D.C. C.

[2]For a brief background on these controls, see John Hewson and Eisuke Sakakibara, "The Impact of U.S. Controls on Capital Outflows on the U.S. Balance of Payments: An Exploratory Study," *IMF Staff Papers,* March 1975, pp. 37-43.

[3]Paul Volcker, Laurie Milbank Centenary Lecture, sponsored by the City University, London, November 27, 1974..

[4]*IMF Annual Report* (Washington, D.C., 1975), p. 40.

[5]*Ibid.,* pp. 23-24.

[6]See, for example, H. Robert Heller, "International Reserves and World-wide Inflation," DM/75/63, International Monetary Fund (Washington, D.C., July 29), 1975.

[7]Gerald A. Pollack, "The Economic Consequences of the Energy Crisis," *Foreign Affairs,* April 1974.

[8]*Oil and International Money* (Transcript of a conference sponsored by Investment and Property Studies, Ltd., London, April 18, 1974), p. 6.

[9]Helen B. Junz, "Some Observations on the Worldwide Intensification of Inflation," in Joint Economic Committee of the U.S. Congress, *Studies in Price Stability and Economic Growth* (Washington, D.C., August 5, 1975), pp. 50-51.

Chapter 8

Oil-Money Recycling and the Transfer Problem*

As discussed in the previous chapters, one of the urgent issues facing the world economy is how to overcome the serious monetary and financial crisis arising from the four-fold increase in petroleum prices imposed in late 1973 and early 1974 by the Organization of Petroleum Exporting Countries. Estimates of the total reserve accumulation by OPEC countries range up to $450 billion by 1980. In order to cope with this serious payments imbalance, a variety of so-called recycling schemes have been proposed. In this chapter, we shall review the essential features of these proposals, and examine their economic meaning in light of the new transfer problem posed by OPEC. The potency of the OPEC cartel has proved to be so persistent and pervasive that a critical review of the major recycling proposals on record is not only timely, but probably will continue to be important for some time to come.

CONCEPTION OF THE ISSUE

The term "recycling" itself has evidently been drawn from the current environmentalist context, in which it is synonymous with — "reusing what would otherwise be discarded." In this sense, to speak of the recycling of oil money is totally misleading, because such money would not be discarded in any event. Instead, if we understand the "recycling" of oil money to mean the transfer of surplus oil revenues from OPEC countries to the rest of the world, then obviously recycling occurs automatically, so long as

*This chapter is co-authored by Prof. Joseph Aschheim, of George Washington University, and myself. An earlier version of this chapter appeared in the November 1975 issue of *Economia Internazionale* as "Oil-Money Recycling: A New Transfer Problem."
Yoon S. Park.

any surplus oil revenues accrue. OPEC countries would not store under the desert sands those American dollars or British pounds that they could not spend on their own imports. Rather, OPEC countries will use their surplus oil revenues for investment either in foreign financial or in foreign physical assets. In other words, oil money not spent on current imports will be accumulated abroad in liquid balances or will be invested abroad directly or indirectly. Thus recycling, in the context of maintaining an accounting balance in the international balance of payments, obtains automatically. This, however, is the trivial meaning of recycling. It is in this trivial sense that William Simon, the U.S. Secretary of the Treasury, has claimed that there is no "recycling problem," while granting that there exists what he has termed a "reshuffling problem."[1]

However, the above definition of "recycling" conveys nothing more than an accounting axiom of international finance, namely that a movement of goods from one country to another will necessarily be offset either by a movement of goods and services in the opposite direction, or by an equivalent financial transfer in the absence of a matching real transfer. We may describe this version of "recycling" as primary or built-in recycling, in that it occurs automatically even without a discretionary recycling plan. In contrast, we may conceive of a second-round recycling in which the surplus oil money, originally invested by oil exporters in one set of oil-importing countries, such as the United States, the United Kingdom and West Germany, is rechannelled to another set of oil-importing countries. In a free-market framework, this second-round recycling need not take place automatically, since those oil-importing countries that do not receive oil-money investments directly from OPEC countries in the primary recycling are mostly poor-credit countries. Therefore, certain non-market arrangements, such as those envisioned in various so-called recycling proposals, may be necessary to achieve secondary recycling.

The fundamental economic problem generated by soaring oil prices is the continuing prospect of massive current-account surpluses of OPEC countries with the rest of the world. Both the magnitude and the persistence of such surpluses are related to the low absorptive capacity of many OPEC countries, in the face of artificially imposed cartel prices for a product with a highly inelastic demand, at least in the short term. In view of this problem, it remains to be seen whether in coming years the international monetary system can continue to function without serious disruption amid huge and persistent oil-induced surpluses. Recycling proposals are intended to ameliorate this situation.

Thus, the immediate issue facing the world economy is how to finance oil imports on terms that have been tightened drastically since the 1973 oil embargo by Arab oil-exporting countries. In principle, the various recycling schemes are concerned with the immediate oil-financing problem, rather than with the long-range question of how the rest of the world would adjust economically to a stream of huge trade deficits vis-a-vis OPEC countries. This long-range economic question has been studied in Chapter 4.

Since the OPEC nations as a group are not likely to invest their surplus oil revenues in each oil-importing country in proportion to its particular oil deficit, some oil-importing countries might experience net reserve inflows, while others might suffer net reserve outflows. The former are likely to be affluent and credit-worthy industrial countries, while the latter are likely to be developing countries as well as some low-credit industrial countries. It is the latter group of countries that will experience an oil-financing problem, for which the various recycling schemes are intended as a solution. Recycling will also help the industrial countries, however, because in the absence of recycling, world trade in general would suffer from a movement towards protectionism by the developing countries. Even without the introduction of protectionism, the industrial countries' volume of exports would still decline under the floating exchange-rate regime, for their currencies would appreciate against the currencies of the latter low-credit group in the absence of recycling. This outcome can be readily illustrated as follows.

Let us divide the world into three groups of countries: Group A for OPEC members; Group B for those oil-importing countries in which surplus oil revenues are invested by OPEC nations; and Group C for the other oil-importing countries in which no oil revenues are invested by OPEC nations. For simplicity, assume that there was initially external balance among A, B, and C. Now inject the oil-price hikes of 1973. After the oil-price upsurge, Group A has a trade surplus with Groups B and C, but A's current account surplus with B and C is offset by A's capital account deficit (A's capital export) with B. Therefore, A has an overall payments balance with the rest of the world (composed of B and C.)

However, even though B and C maintain current-account balance with each other, B now has an overall payments surplus because of capital inflows from A in excess of B's own oil payments to A. On the other hand, C has an overall payments deficit, because C's oil deficit is not offset by a comparable capital inflow from A. Due to the limited absorptive capacity of A, the only way that C can recover from its trade deficit with A is to incur a payments surplus with B in the amount of C's deficit with A, which corresponds exactly to B's payments surplus. Discretionary recycling, by transferring to C those capital inflows into B from A which are in excess of the increase in B's oil payments, is one way of achieving this end. In the absence of such recycling, the only other alternative is for B to incur a trade deficit with C, an outcome achieved automatically through the appreciation of B's currencies against C's currencies within the floating exchange-rate regime. However B will then have trade deficits on two fronts—with A due to oil imports, and with C due to the offsetting of C's oil deficits. Of course, B will in all likelihood be most reluctant to run trade deficits on both fronts. It is troublesome enough for B to have a huge oil deficit with A, without also running another trade deficit with C. Therefore, recycling appears to be in the interest of B as well as of C, as long as we assume that the funds recycled to C will be duly repaid in the future.

Originally, the B group might have thought that recycling was only for the C Group to worry about; for it is the C Group that has an oil-financing problem, and recycling is principally concerned with the oil-financing problem rather than the basic trade-balance issue. For example, the United States, which is in the B Group, initially tended to downgrade the importance of recycling, while putting primary emphasis on a reduction of oil prices as its preferred solution. However, the American attitude subsequently changed, as evidenced by the formulation of the Kissinger-Simon Plan for recycling~~, to be reviewed below~~. This change might have derived in part from a new recognition of the economic implications for the United States of the absence of recycling, so long as the OPEC pricing policy persists.

Basically, there are two ways to accomplish recycling. Since the purpose of recycling is to transfer funds to the C Group sufficient to cover its oil deficit with the A Group, recycling can be achieved either by the direct transfer of surplus oil revenues from A to C (primary recycling), or by the indirect approach of rechannelling those surplus oil revenues already invested in B into C (secondary recycling). The various oil-money recycling schemes involve either primary recycling or secondary recycling, or a mixture of both.

All recycling schemes are intended to facilitate oil-money recycling to Group C from either Group A or Group B, neither of which would be willing to lend surplus revenues to Group C in a free-market framework. In contrast to such *discretionary* (artificial) recycling in the form of primary or secondary recycling, we previously described the *automatic* primary recycling from Group A to Group B. This recycling is automatic in the sense that Group A voluntarily invests its surplus oil revenues in Group B without the help of recycling schemes, and such recycling is also primary recycling because the oil revenues are recycled directly out of Group A. It may be asked whether there can be an automatic secondary recycling. Since it is likely that Group C will also be considered credit-deficient by Group B (as it is by Group A), secondary recycling will not be automatic but discretionary.

To sum up, we can divide recycling into primary and secondary recycling, depending upon whether oil money is recycled out of

Group A directly or out of Group B indirectly. Additionally, recycling can be either automatic or discretionary, depending upon whether the particular recycling takes place automatically in a free-market framework or artificially with the help of recycling schemes. We have also seen that primary recycling can be either automatic or discretionary, but that secondary recyling is always discretionary. The above relationships may be expressed in the following tabulation:

	Primary recycling	Secondary recycling
Automatic recycling:	Group A to Group B	(Not applicable)
Discretionary recycling:	Group A to Group C	Group B to Group C

It has been argued by some that a recycling scheme should consider not just the narrow aspect of the *oil*-financing problem, but the general issue of financing payments deficits throughout the world arising from price increases of various commodities such as copper, coffee, wheat, aluminum, etc. While acknowledging the merit of this point, we must confine our focus to the oil-financing problem alone. In what follows, therefore, we limit ourselves to considering the economic issues involved in the attempts to respond to the need for financing huge oil deficits.

From the foregoing analysis, we infer that recycling may affect not only the financing of oil imports by overall-deficit countries, but also the restoration of external balance for those oil-importing countries with overall surpluses. In this connection, it has recently been suggested that the economic implications of recycling can be viewed within the framework of the "transfer problem."[2] Let us consider this suggestion.

THE NEW TRANSFER PROBLEM

The subject of the German transfer problem sparked lively discussions in the economics literature of the interwar period, involving such authorities as Hjalmar Schacht, Fritz Machlup, John Maynard Keynes, Bertil Ohlin, Jacques Rueff, Gottfried

Haberler, and others. As Machlup has responded to the more recent claims that the transfer problem is a "dead issue" of merely historical interest, "the transfer problem is alive and kicking; indeed, it is one of the most troublesome economic problems of our days."[3]

The immediate context in which the transfer problem came to be identified as such was the issue of German reparation payments after World War I. In that context, the transfer problem was essentially based on the following question: if a country undertakes to pay large war indemnities abroad, will it thereby exhaust its external reserves, or will its trade balance adapt to the outflow of funds by generating an offsetting funds inflow with which the indemnification commitment can be met? The assumed exchange-rate system within which this question was considered was that of pegged or fixed exchange rates, subject to variation only as a last resort.

Logically, there is no reason to limit the concept of a transfer problem to the case of war-reparation payments, or to apply this concept only within the context of a fixed exchange-rate system. In other words, any undertaking by a country to make large payments abroad, for whatever reason, involves a transfer problem. And in any such case, the practice of foreign-exchange control or, alternatively, the adherence to a floating exchange-rate system would, at most, only suppress or disguise the existence of a transfer problem; it would not prevent the emergence of the problem.

Originally, the transfer problem was envisaged as a dual one involving both the budgetary task of extracting the necessary sums of money from taxpayers, and the transfer task of converting the domestic funds so levied into the foreign currency with which the debt obligation is to be settled. Accordingly, Machlup is enlightening when he points out that from 1946 to 1958 the United States made heavy foreign remittances but that no "transfer difficulties" were then felt.[4] The year 1958, he indicates, marked a turning point in that some central banks that had been accumulating dollar reserves began converting them into gold; from that point onward, the United States began to experience a transfer problem.[5]

Writing just before the large-scale American involvement in Vietnam commenced, Machlup made a comparison of the transfer problems in four countries: Britain in 1793-1816, France in 1871-1875, Germany in 1924-1932, and the United States in 1950-1963. In comparing these four cases, Machlup surprisingly found that the "burden on the economy of the paying country was in several respects least heavy in the case of Germany."[6] An even more remarkable conclusion of Machlup's was his "highly astonishing finding that the burden of foreign payment is so very heavy in the case of the United States."[7] We should not find fault with him that just *before* the onset of the American commitment to the war in Vietnam, Machlup was willing to advance the prognosis that "the transfer difficulties of the United States will be solved"[8] through various developments that might well have materialized as he had forecast, had it not been for the economic magnitude of American involvement in the war in Vietnam. Nor can we blame Machlup for not having anticipated the present transfer problem caused by the massive oil-price upsurge emanating from the OPEC cartel policy; for who did?

In paying for oil imports, the United States and, to a limited extent, the United Kingdom have no transfer problem at least for now, because their currencies are also the transactions currencies for the oil trade. On the other hand, other oil-importing countries are faced with the transfer problem of converting their domestic currencies into the oil-transaction currencies. While the German reparations obligations constituted a non-trade-caused transfer problem that called for solution by adjustment through trade accounts, oil-money recycling is a trade-caused transfer problem to be solved by non-trade (capital-account) adjustment, at least in the short term. Due to the limited absorptive capacity of the oil-exporting countries, attempts to cope with recycling through trade adjustments would only lead to severe dislocations in the world economy.

Thus, there is a contrast between the fiscal task in the old reparations-type transfer problem and the fiscal task in the new oil-induced transfer problem. In the former case, the fiscal task involved a deflationary domestic policy to effect a trade-balance surplus with which to engender a transfer-currency inflow. In the latter case, however, oil-importing countries are faced with the task of reflating their domestic economy to compensate for the diminished propensity to consume as a result of the "excise-tax" effect of the increase in oil prices.

The current oil-induced transfer problem is also different in a global sense from the traditional transfer problem. The oil-importing countries as a group do not need to convert their currencies into the currencies of oil-exporting countries to make the transfer, because the payment currencies for oil imports are actually the currencies of certain importing countries. Thus, there is no traditional transfer problem of converting the transferor's currency into that of the transferee. As a result, while the traditional transfer problem required the cooperation of the transferee in order for a transferor to execute the transfer, e.g., by allowing the latter to borrow from, or export to, the former, the new transfer problem does not absolutely require the cooperation of oil-exporting countries. The new transfer problem can be solved by the oil-importing transferor countries themselves. As we noted previously, however, an intra-group solution of the new transfer problem by the oil-importing countries would require those countries with an overall payments surplus (due to capital inflows from oil exporters in excess of the increased oil import bills) to extend enough credit to the other oil-importing countries to cover their net oil deficits.

But some of the deficit oil-importing countries may not have the necessary credit standing to obtain credit from the surplus oil-importing countries in a free-market framework. Even those deficit oil-importing countries that do have the necessary credit standing may still not be able to obtain credit in sufficient volume to cover their oil deficits, because the amount involved may be beyond the capacity of private financial intermediaries. Here enter the secondary recycling schemes, whose aim is to facilitate the necessary credit extension from surplus oil-importing countries to deficit oil-importing countries which would be impossible in a free-market framework.

Strictly speaking, a transfer problem involves converting accumulated domestic funds into a foreign-transaction currency. In this sense, only the United States of all the oil-importing countries appears not to have a transfer problem, because the U.S. dollar alone is the transaction currency in the oil trade.[9] If the United States were willing to supply the necessary dollars to all deficit oil-importing countries, there would be no transfer problem in the current oil crisis. However, the United States would be extremely reluctant to fill by itself the role of international financial intermediary to all poor-credit oil-importing countries. A similar reluctance may also arise in other credit-worthy oil-importing countries, such as Germany and Switzerland, that will receive surplus oil money as investments. Yet, when the United States and other credit-worthy oil-importing countries experience inflows of surplus oil revenues (beyond their own oil-import bills), their currencies will appreciate under the floating exchange-rate regime, and their exports will diminish correspondingly. In order to avoid currency appreciation while at the same time avoiding the role of intermediary to poor-credit countries, credit-worthy oil-importing countries may discourage, or simply refuse to accept, investments from OPEC members.

In early 1975, Switzerland imposed a 40% annual tax on foreign Swiss franc deposits at Swiss banks in order to discourage capital inflows, many of which presumably came from OPEC members. The Swiss government took this measure to protect its export industries and the tourist trade by preventing further appreciation of the Swiss franc. If all other credit-worthy countries had taken similar measures, OPEC investors would have been forced to invest exclusively in the Eurocurrency market. However, some Eurobankers are already raising the possibility of negative interest levies on OPEC deposits like those in Switzerland, since some Eurobanks are near their limit in accepting large OPEC deposits.[10] Certainly, the private Eurobanking system would not be able to cope with the huge oil-money deposits that would thus be forced out of credit-worthy oil-importing countries.

Hence, OPEC members have several alternatives. First, they may cut back their oil production to the level necessary to cover only their current foreign exchange requirements, such as imports, foreign aid, military needs, etc. Second, they may behave like mercantilists, purchasing and hoarding gold on a large scale. And third, they may invest surplus oil revenues in poor-credit oil-importing countries. The first alternative involves a rather high risk of economic (and perhaps even the threat of military) retaliation by oil-importing countries, because such a cutback in oil production would be tantamount to economic injury to the oil-importing countries, rich as well as poor. The second alternative is also unattractive to OPEC members, because the world gold price would soar, largely to the benefit of non-OPEC gold-producing countries. In the third alternative, OPEC members would be assuming the high credit risks of poor oil-importing countries that may seek to borrow large sums of money.

In these circumstances, OPEC countries might adopt a combination of all three alternatives: that is, some oil production cut-back, some gold purchase, and some investments in poor-credit countries. However, it will be more advantageous to OPEC if both OPEC members *and* credit-worthy oil-importing countries cooperate in solving the recycling problem by shouldering the credit risks of poor oil-importing countries on a joint basis. Some of the recycling plans which have been proposed are based on this principle of cooperation between oil-exporting and oil-importing countries. The following sections review these cooperative recycling schemes.

PRIMARY RECYCLING SCHEMES

As noted earlier, some recycling takes place automatically. It turns out, however, that this automatic recycling benefits only a limted number of rich industrial countries. Because these countries have well-developed financial and non-financial markets offering attractive investment opportunities, surplus oil money tends to flow to them, even though they could manage without such capital inflows, while avoiding poor developing countries that lack such funds to begin with. Therefore, proposals for primary recycling aim at changing the *direction* of this initial, automatic recycling. Since automatic primary recycling is not likely to benefit those poor oil-importing countries that are most in need of external funds, all primary recycling schemes propose to set up intermediary institutions to channel oil-money flows from OPEC countries directly into needy countries.

The most widely noted proposal among the primary recycling schemes is the Healey Plan, suggested by the Chancellor of the Exchequer of Great Britain at the 1974 annual meeting of the IMF and the World Bank. Under this plan, the IMF would expand its present oil facility by borrowing about $30 billion per annum from oil-producing countries at market interest rates. The IMF would then re-lend the money to these oil-importing countries in need of external funds, according to the lending procedures of the IMF oil facility. As there is no certainty that the $30 billion figure is the optimum amount, the figure can easily be increased or decreased according to the acknowledged need. The basic features of the Healey Plan are that the IMF would assume the main intermediary role, and that both developing and developed countries could borrow from the facility to cover their oil deficits. In January 1975, the IMF Interim Committee agreed to adopt a modified version of the Healey Plan, with its size reduced to $6 billion, as an extension of the 1974 IMF oil facility. On April 4, 1975, the IMF formally launched the scheme as the 1975 IMF oil facility.

Several objections have been raised to the Healey Plan. First of all, it does not provide concessionary rates for developing countries, and treats developing and developed countries alike as long as they incur oil deficits. However, provisions for interest subsidy to developing countries could be incorporated into the Healey Plan without much difficulty. The IMF Interim Committee adopted an interest subsidy plan in association with the modified Healey Plan. For this purpose, it was agreed to establish a Special Account with contributions from oil-exporting and industrial nations.

More fundamental objections revolve around the issue of loan guarantees and the so-called "unwieldy" membership of the IMF. If a country were to default on its loan from the IMF oil facility, rich industrial countries like the United States and Germany would be liable for a substantial portion of the defaulted loan, since their IMF quotas are among the highest. Since the Healey Plan would be managed by an international organization largely insulated from their direct influence, national govenments would be reluctant to shoulder a defaulted loan made under the Healey Plan. Furthermore, the oil-producing countries would presumably be unwilling to entrust a large portion of their surplus oil revenues to the IMF, in which the Western industrial countries together have a far greater voice than do the OPEC countries. It is a well-known fact that OPEC members have been rather cautious in their contributions even to the much smaller IMF (Witteveen) oil facility of 1974. It appears doubtful whether OPEC nations would contribute tens of billions of dollars as envisaged in the Healey Plan.

Another primary recycling proposal has been provided by a group of five economists with diverse banking and industrial connections, in an article in the January 1975 issue of *Foreign Affairs.*[11] They advocate setting up two different facilities, an OPEC Fund for Government Securities and an OPEC Mutual Investment Trust(s), in which OPEC countries would have a dominant (or majority) voice by contributing all the capital resources. The Fund for Government Securities would invest in government securities exclusively, and the Mutual Investment Trust would make investments in private enterprises and joint ventures.

This proposal, however, is deficient as a meaningful recycling scheme, because it makes no provisions for a guarantee against defaults by "poor credit" countries. In consequence, both the Fund and the Trust might channel OPEC investment funds to the same rich and credit-worthy countries that OPEC countries would have chosen by themselves without this plan. A primary recycling scheme is meaningful as a rechannelling mechanism only when its existence helps redirect the original flow of oil money from rich countries toward poor and creditless countries that would suffer oil-induced deficits in the absence of discretionary recycling. All that both the Fund and the Trust would accomplish would be the interposition of an added party, a middleman between OPEC countries and the rich oil-importing countries that would be receiving surplus oil money even without the middleman. Therefore, the foregoing proposal appears to be principally a business proposition, which would appeal primarily to those financial institutions that are most apt to engage in investment advisory services for oil-producing countries. Unless rich oil-importing countries were required to shoulder the default risks of poor oil-importing countries, OPEC countries would not be induced to invest their funds in poor countries in the form of either government securities or private ventures. Both the Fund and the Trust(s) would be unable to genuinely rechannel oil revenues to countries which could not obtain the needed funds on their own credit.

Another primary recycling scheme, the Fund for Capital Recycling, has been proposed by George W. Ball.[12] The Fund, in which OPEC nations would have a co-equal voice with the Western industrial nations, would receive investment funds from OPEC nations and invest them in government securities that might be patterned after British Consols. However, unlike the scheme proposed by the five economists, Ball does suggest a way in which a meaningful recycling can be achieved for creditless countries. In order to help less developed countries, the Fund for Capital Recycling would establish concessionary lending rates, subsidized equally by OPEC nations and industrial countries, for its loans to LDCs. LDCs would then be in a strengthened position to service their oil-money borrowings. In this fashion, the Fund could redirect some of the primary recycling from OPEC countries to LDCs. Presumably the default risk for the loans to LDCs would be shouldered jointly by the OPEC countries and the Western industrial nations.

There have been other interest-subsidy proposals, according to which surplus funds of OPEC countries would be lent to LDCs at concessionary interest rates, through contributions from rich countries for the difference between the concessionary rates and market rates. One such plan was proposed by the Trilateral Commission, composed of private citizens from North America, Western Europe, and Japan. It calls for the World Bank to borrow funds from OPEC countries at market interest rates and lend them to LDCs at 3 percent through a separate "lending window" of the World Bank. The difference in interest cost would be subsidized by industrial countries. A similar plan has been proposed by Pedro-Pablo Kuczynski, of Kuhn, Loeb & Company, a New York investment bank.[13] Under this plan, the IMF would borrow funds from OPEC countries at market rates and lend them to LDCs at concessionary rates. The cost of interest subsidy would be borne by OPEC countries.

A rather unique primary recycling scheme has been proposed by Walter Salomon, Chairman of Rea Brothers, an investment bank.[14] Under this plan, oil-importing countries would now pay in cash only the amount that would have been due at the pre-October 1973 oil prices, with some adjustment for inflation. The difference between the old and the new prices would be paid in 3.5%-per-annum 25-year bonds, whose trusteeship would repose in the Bank for International Settlements with an overall limit of $300 billion. The OPEC members would be allowed to tender these bonds in payment for the importation of goods and services from oil-importing countries, but not in payment for the acquisition of assets in the latter countries. Salomon notes that a similar plan has already been adopted by one of the OPEC members, Venezuela, which has reached an agreement with Costa Rica, El Salvador, Guatemala, Honduras, Nicaragua, and Panama under which these countries are to import oil from Venezuela at market prices ($11.90 per barrel), but $6 a barrel will be handed back as a five-year loan at 8 percent. The main drawback of this proposal, however, is that OPEC members as a group are quite unlikely to accept it. It is also doubtful that the bonds will be accepted by traders as payment for imports of OPEC countries.

SECONDARY RECYCLING SCHEMES

As previously noted, secondary recycling schemes are intended to redirect surplus oil money already invested in one group of oil-importing countries into the other group of poor (or creditless) oil-importing countries. The best known proposal among them is the Kissinger-Simon Plan. The gist of the Plan is as follows: let the OPEC countries invest their surplus oil revenues in the first instance wherever they choose. Industrial countries would then establish, in association with the OECD, a $25 billion supplementary loan facility, which would raise the money from the capital market, or from the governments, of the oil-money-receiving countries. The money thus raised would be lent to countries participating in the Plan in need of supplementary loans to finance their oil-related or even overall payments deficits. All member countries of the facility would share the credit risk of the loans to members in proportion to their share of participation, to be decided on the basis of their respective oil import bills, GNP, and international trade.

The main objection to the Kissinger-Simon Plan has come from members of the European Community that do not favor the preconditions for a loan. The United States has proposed that the loan facility (or the so-called solidarity fund) should lend only to those member countries that would assume the dual commitment to cooperate in reducing dependence on oil imports and to follow other responsible adjustment policies. The United States has also proposed that decisions on the provision of financial support should be made by a weighted voting system for participants, in which the United States would have the largest vote. These preconditions are viewed by some European countries as an attempt by the United States to intervene in their internal affairs. Furthermore, countries with a strong payments position, such as West Germany, are apprehensive that they would be unduly burdened by large cash contributions or guarantees for loans in bailing out creditless oil-importing countries. They believe that a recycling scheme should be based upon cooperation and burden-sharing among both oil-exporting and oil-importing countries. Since the Kissinger-Simon Plan would exclude oil-exporting countries, only those oil-importing countries that have a strong payments position would be unduly burdened in solving a problem which should be a global responsibility. Also, the Plan does not benefit those LDCs—perhaps most—that may not be allowed to participate in the OECD-centered scheme.

In April 1975, the Kissinger-Simon Plan was agreed upon by the ministers of OECD member countries. Under the agreement, the OECD members will jointly set up a Financial Support Fund (FSF) of SDR 20 billion (about $25 billion). For an initial period of two years, the FSF will grant loans to its member countries in payments difficulties arising mainly, but not exclusively, from oil imports. There will be two main methods for the FSF to raise the necessary funds. The first is through individual commitments under which each member country called upon provides either funds directly to the FSF or an individual guarantee for borrowing by the FSF in financial markets. The other method is through borrowing by the FSF in financial markets on the basis of a collective guarantee by all members.

Another secondary recycling plan, initially proposed by the Secretary General of the OECD, provides that the Bank for International Settlements borrow money in the international capital market and re-lend it to oil-deficit nations. It is presumed that a substantial portion of surplus oil money would be invested by OPEC countries in the international capital market, where the BIS would borrow by issuing bonds. Here too the issue that has been raised is how the BIS loans could be guaranteed; some feel that guarantees for the BIS loans should be undertaken by central banks of the major industrial countries that compose the BIS's Board of Directors.

COMBINING PRIMARY AND SECONDARY RECYCLING

There is no *a priori* reason why both OPEC countries and rich industrial countries receiving oil money could not jointly contribute to a lending facility that would lend to countries experiencing difficulties in financing their oil imports. One such plan has been proposed as a part of the Kissinger-Simon Plan. While the solidarity fund would be limited to industrial oil-importing countries, the United States has also proposed to establish a Trust Fund, to be managed by the IMF. Both OPEC and industrial countries would contribute funds to the Trust Fund, whose resources would further be augmented by a part of the profits derived from the sale of a small portion of the IMF's gold holdings. The Trust Fund would offer credit to the poorest developing countries on soft terms—perhaps 2-4 percent interest and moderately long maturities.

Another such recycling scheme is the European Community loan. In September 1974, the Council of the European Communities agreed to provide assistance to some member countries in financing deficits caused by the increase in oil prices. Under this plan, which is expected to be in operation before 1976, the EEC Council will borrow both from OPEC countries and on the world capital market, and then re-lend the funds to member countries in payments difficulties. If any borrower defaults, the other members of the European Community will be obligated to repay the loan in proportion to their wealth.

SOME POLICY INFERENCES

Oil revenues would be recycled automatically in a global sense even without a recycling scheme. A recycling plan is meaningful only when it is designed to change the direction of automatic recycling from rich oil-importing countries to poor and creditless oil-importing countries. According to this criterion, the proposal for the OPEC Fund for Government Securities and the OPEC Mutual Investment Trust(s) does not constitute a recycling scheme as such. Recycling is in the joint interest of both rich and poor oil-importing countries, and of oil-exporting countries as well. In the absence of recycling, poor and credit-deficient oil-importing countries would be more willing to resort to import restrictions and beggar-thy-neighbor policies. Additionally, the currencies of rich oil-importing countries receiving surplus oil money would appreciate vis-a-vis the currencies of poor oil-importing countries. As a result, the exports of rich oil-importing countries to poor and credit-deficient oil-importing countries would decline and perhaps create deficits for them. This situation could become unacceptable to rich oil-importing countries, because they would suffer trade deficits with both oil-exporting countries and poor oil-importing countries. In this predicament, rich oil-importing countries operating within the floating exchange-rate regime might discourage, by an outright ban or by imposing low or negative interest rates, inflows of surplus oil money from OPEC countries. The latter countries, in turn, would then be compelled to invest their surplus oil money in less than prime-credit oil-importing countries.

In this manner, a market mechanism in the context of the floating exchange-rate regime would work to help achieve automatic recycling among financially sound oil-importing countries even in the absence of a specific recycling plan. However, such automatic recycling through a market mechanism may not help many less developed countries which, if left to themselves, would have no prospect of attracting enough oil-money investments to meet their external-credit needs. These countries perceive a real and urgent need for a recycling plan that would not only rechannel oil money to them, but would do so at concessionary interest rates. Thus, it may be advisable to have two separate recycling schemes: one for industrial countries in combination with those developing countries that are progressing rapidly, and another for the poorest countries. The first recycling scheme could take the form of a multilateral guarantee or insurance policy for the loans being raised by individual countries either directly from OPEC members or on major capital markets of the world. In other words, such a recycling scheme would not require its own special fund; it could exist merely in the form of a multilateral insurance policy. On the other hand, the second recycling scheme would require an independent lending facility, whose resources would be supplied by both OPEC nations and those industrial countries receiving surplus oil revenues. Loans would be made on concessionary terms with respect to both interest rates and maturity. The lending facility could be established in an international institution having a broad membership, such as the IMF or the World Bank. The cost of interest subsidy might be borne by contributions from both OPEC and industrial countries. As an alternative to possible upheaval in the world economy in the absence of such recycling, the cost of an interest subsidy would appear to be a modest price for both OPEC and industrial countries to pay.

Finally, we turn to difference in population size as an important factor in the absorptive capacity of different national economies. Thus, Kuwait, Abu Dhabi, and Qatar, Persian Gulf states with huge oil reserves and relatively few people, are dwarfed in their absorptive capacity by Egypt, with its relatively meager oil reserves and massive populace. Hence, the historical process underlying the manner in which present national boundaries came to be drawn among former colonial possessions contributed significantly to the magnitude of the new transfer problem triggered by OPEC. Great Britain, which, as the last colonial power, had carved up portions of the former Ottoman Empire after World War I, played the decisive role in creating the boundaries between the now independent Persian Gulf states.

In this vein it has been noted, not altogether facetiously, that the American nation is fortunate that Thomas Jefferson and Abraham Lincoln bought out the French and Russian territories in North America before the importance of petroleum became established; "otherwise the French might have set up a series of oil mini-states in the Gulf of Mexico and the Russians might have matched them in Alaska. Today such mini-states might be independent voting members of the United Nations with their separate rulers speculating in the international money markets and controlling oil supplies, while Illinois might be importing oil from Indonesia."[15] The point is that vast disparities in percapita oil wealth may persist even among neighboring countries whose populations share the same ethnic and cultural heritage.

Accordingly, one solution to the oil-induced transfer problem entails the re-drawing of some national boundaries to integrate nation states of the same population stock. It can be taken for granted that, in the current, post-colonial era, any boundary redrawing under peacetime conditions would have to be undertaken with the joint consent of the parties. That generating such consent is no simple matter is amply illustrated by the aborted 1973 plan for the merger of Libya, a sparsely populated major oil-exporting state, with its populous neighbor Egypt. Yet, such a merger would be mutually beneficial in terms of the complementarity of the resource endowment of the two countries, as shown by the application of the Heckscher-Ohlin theorem of international differences in comparative advantage. All the more commendable would such a merger be when viewed in the context of the ethnic and cultural kinship of the populations of the two countries and the contiguity of their territories. Thus, a subject that warrants future economic, as well as political, inquiry is the extent to which OPEC's price policy may yet stimulate the redrawing of national frontiers between some oil-rich and some oil-poor countries in years to come. In the context of present political realities, of course, it would be highly unrealistic to expect any immediate trend toward such mergers; this would take place only when their potential advantage, in the form of the increased political security gained by belonging to a bigger and more powerful country, were to outweigh the potential disadvantage of losing the economic comforts afforded by oil revenues.

NOTES

[1]For criticisms of such a simplistic view on generalized reshuffling, see Otmar Emminger, "International Financial Markets and the Recycling of Petrodollars," *The World Today,* March 1975, p. 98.

[2]See, for example, Jan Tumlir, "Oil Payments and Oil Debt in the World Economy," *Lloyds Bank Review,* July 1974.

[3]Fritz Machlup, *International Payments, Debts, and Gold* (New York: Charles Scribner's Sons, 1964), p. 368.

[4]*Ibid,* p. 390.

[5]*Ibid,* p. 391.

[6]*Ibid,* p. 392.

[7]*Ibid,* p. 393.

[8]*Ibid,* p. 393

[9]The British pound is gradually losing the role of transaction currency. While there have been sporadic talks among some OPEC members of replacing the dollar by a composite currency unit as the new oil transaction currency, the present analysis applies until such action is taken.

[10]See, for example, "Tax on Petrodollar Deposits?" *International Insider,* January 27, 1975.

[11]K. Farmanfarmaian, A. Gutowski, S. Okita, R. Roosa, and C. Wilson, "How Can the World Afford OPEC Oil?" *Foreign Affairs,* January 1975.

[12]Comments by George W. Ball, Senior Partner, Lehman Brothers, at the Fiftieth Anniversary Conference of the Bankers Trust Company, London, September 17, 1974.

[13]Pedro-Pablo Kuczynski, "Recycling Petrodollars to the Third World—II," *Euromoney,* December, 1974.

[14]Walter H. Salomon, "Bonds for the Arab Oil Funds," *The (London) Times,* January 29, 1975.

[15]From correspondence by Harry J. Lipkin of the Weizmann Institute of Science, Israel.

Chapter 9
The Oil Crisis and Developing Countries

The effect of oil-price increases on developing countries with no oil resources has been far more severe than on industrial countries for several reasons. First, non-oil developing countries have relatively little room for cutting down on oil imports through energy conservation, because most of their oil consumption is for essential productive purposes. Second, the unprecedented rise in the cost of oil has deeply affected the production of fertilizers. Some developing countries are having difficulty in getting adequate supplies of fertilizers, and thus are forced to reduce food production. Third, the terms of trade have moved strongly against non-oil LDCs due to the twin factors of the oil-price hike and the subsequent rise in the import prices of manufactured goods from developed countries. The World Bank has studied 40 non-oil LDCs and found an average decline of 15 percent in their terms of trade between 1973 and early 1975.

In contrast, many industrial countries have been able to compensate for much of the rise in oil prices by increasing their industrial export prices much faster than the oil content of their exports has increased in cost. The result has been that the burden of the OPEC trade surplus falls more heavily on non-oil developing countries than on industrial countries. The dollar value of world trade grew by 45 percent in 1974; 18 percent of this rise was caused by the increase in oil prices, and 13.5 percent was accounted for by the inflated prices of manufactured goods from industrial countries. Only 9 percent of the increased trade value came from the higher prices of commodities exported from non-oil LDCs.

As a result, the world trade share of non-oil LDCs decreased from 15.5 percent in 1970 to 14.3 percent in 1974.

Table 9.1
Shares of Developing Countries in World Trade, 1950-70 (in %; based on current US dollar values of exports)

	1950	1960	1970	1974
All LDCs of which:	35.4	26.2	22.2	31.4
(1) Oil-exporting countries*	7.0	7.0	6.3	16.8
(2) Middle income LDCs	22.4	15.1	13.2	12.9
(3) Lower income LDCs	5.4	3.6	2.3	1.4

*Including several small oil-exporting developing countries which are not OPEC members.

Source: *Prospect for Developing Countries: 1976-1980,* International Bank for Reconstruction and Development, July 8, 1975, p. 47.

The plight of non-oil LDCs became exacerbated further in 1974 and 1975 as their volume of exports declined significantly due to the serious recession in the industrialized countries. According to an estimate by the OECD Secretariat, in 1974 and early 1975 the exports of non-oil LDCs to the Western industrial countries were cut by $5.5 billion from what they would have been without the slump in demand from the rich countries, and those to the centrally-planned (Communist) countries declined by perhaps $1.5 billion. In contrast, Western industrial countries have been able to score a dramatic recovery in their trade balances due to their economic recession as well as a sharp increase in OPEC imports from them. From a combined trade deficit of $6 billion in the first half of 1974, the top ten industrial countries in the West moved to a combined deficit of $0.9 billion in the second half of 1974, then to a combined surplus estimated at $18 billion in the first half of 1975.

The abrupt rise in oil prices directly added about $8.5 billion to the import bills of non-oil LDCs in 1974, and consequently they

Table 9.2
Current-Account Balances of Non-OPEC LDCs
(millions of U.S. dollars)

	1974	1975
Brazil	-6,981	-5,300
Mexico	-2,613	-3,050
South Korea	-1,879	-2,200
India	-1,400	-1,600
Pakistan	-1,039	-1,600
Peru	- 685	-1,125
Malaysia	- 373	- 950
Argentina	245	- 950
Philippines	- 280	- 825
Taiwan	- 825	- 300
All others	-11,970	-17,100
Total	-27,800	-35,000

Source: *World Financial Markets,* Morgan Guaranty Trust Co. of New York, October 20, 1975, p. 5.

Table 9.3
Trade Balances of the Top 10 Nations (billions of U.S. dollars)

	Second Half 1974	First Half 1975*
United States	-3.8	5.4
Canada	-0.1	-0.3
Britain	-6.2	-2.7
Germany	10.5	10.8
France	-1.4	1.5
Japan	2.7	4.0
Italy	-2.4	-0.9
Netherlands	0.2	0.3
Belgium	0.6	1.0
Switzerland	-1.0	-0.7
10-nation balance	-0.9	18.4

*Estimated

Source: *Monthly Economic Letter,* First National City Bank, New York, August 1975.

Table 9.4
Current-Account Balances of Major Groups of Countries* Excluding Official Grants (billions of U.S. dollars)

	1974	1975	1976
OPEC countries	+65	+36	+34
Group of Ten	-12	+16	+ 4
Other industrial countries	-16	-16	-14
Non-OPEC LDCs	-28	-35	-28

*Balances shown do not add to zero because of (1) asymmetries arising from coverage, classification, timing and valuation in reporting of individual countries and (2) socialist and other countries not included in four groups shown.
Source: *World Financial Markets,* Morgan Guaranty Trust Co. of New York, October 20, 1975, p. 5.

suffered a combined trade deficit of $28 billion in the year. Because of the continuing recession in the West, non-oil LDCs are expected to suffer an even worse trade deficit in 1975. It is estimated at $35 billion by Morgan Guaranty Trust Company of New York, while the United Nations Conference on Trade and Development (UNCTAD) estimates the combined 1975 trade deficit of non-oil LDCs at between $33.5 and $42 billion.[1] Developing countries in general have only a limited range of potential export goods that can be raised in volume significantly in the near future. Furthermore, non-oil LDCs are dependent on oil for well over half of their energy, whereas the developed countries get less than half their energy from this source. Consequently, non-oil LDCs are compelled to borrow abroad to finance the increased trade deficit caused by oil price hikes. In 1974, non-oil LDCs received foreign capital inflows of $34.4 billion from both official and private sources, compared to an average foreign capital inflow of $15.4 billion per year during the period from 1969 through 1973. In 1975, non-oil LDCs are expected to further increase their dependence upon foreign capital by borrowing about $43 billion from both private and official lenders.

Beyond these general effects on all of the non-OPEC developing countries, however, the impact of oil-price increases varies greatly among individual developing countries. We may divide the non-OPEC LDCs into three groups according to the severity of the oil-price impact on them. [2] A first group of countries consists of those non-OPEC LDCs that, on balance, have not significantly suffered from the price trends of the past few years and those that may have even benefited from the general price developments. Some of these countries are self-sufficient in oil or export some oil (China, Columbia, Mexico, Peru, Bolivia, and Tunisia); some benefit substantially from their exports of other raw materials whose prices have been increasing (Malaysia, Morocco, Brazil, Bolivia, and Tunisia).

A second group of non-OPEC LDCs consists of the countries which have suffered heavily from the economic recession in the industrial countries as well as from the oil-price increases. Countries like South Korea, Hong Kong, Taiwan, and Singapore have been significantly dependent upon their exports to the industrialized countries in developing their economies. In addition, Greece, Turkey, Spain and Yugoslavia used to benefit from their substantial foreign exchange earnings from tourism and workers' remittances, as well as from exports to the industrial world. The economies of these developing countries are affected adversely by worldwide inflation, economic recession, and sharp increases in the prices of raw materials, including oil.

A third category of countries comprises the hard core of seriously troubled LDCs, totaling about forty in number with a combined population of almost one billion. Most of these countries are located in tropical Africa, South Asia, and the Central American-Caribbean area. For this group of countries, the recent series of economic developments, caused directly and indirectly by the oil crisis, has been overwhelmingly negative. Most of these countries not only are the poorest in the world, but also have the most dismal growth prospects for the future.

For those non-oil LDCs belonging to the second and third categories, the immediate need is to find enough foreign exchanges in order to meet their critical external capital requirements. As we have observed in Chapter 6, however, private lending institutions in the Western industrial countries are increasingly reluctant to extend credits to most non-oil LDCs belonging to the second and third categories. The Euromarkets are still open to a limited number of creditworthy non-oil LDCs, especially those belonging to the first category. But even these countries are not likely to be able to tap Euroloans on the scale necessary to meet all of their external capital requirements. During the first nine months of 1975, non-OPEC LDCs borrowed $6,023 million of Eurocredits, accounting for 40 percent of the total publicized Eurocredits during the first three quarters, compared to $5,844 million, or 20 percent, for the entire year of 1974. However, 80 percent of the $6,023 million was borrowed by just six developing countries (Brazil, Mexico, Hong Kong, South Korea, Phillippines, and Malaysia). This indicates that the range of non-OPEC LDCs to whom Eurofinance is available is fairly narrow.[3] The following tables (Tables 9.5 through 9.9) show the accounts of Eurocredits borrowed by various countries from 1970 through 1974.

The Eurobond market is even less accessible to LDCs, because, unlike the more sophisticated Eurobanks which extend Eurocredits, Eurobond investors are less able to evaluate the country risk of an LDC and thus tend to stay away from Eurobonds issued by an LDC. In 1974, LDCs were responsible for some 10 percent of all borrowing in the Eurobond market; in the first half of 1975, their share declined to about 5 percent. In the Euromarkets, the key question being asked now is whether the current batch of outstanding loans to the LDCs will be successfully repaid.[4]

Table 9.5
Publicly Announced Eurocredits to Non-oil Producing LDCs (in millions of dollars)

	1974	1973	1972	1971	1970	Total 1970-1974
Argentina	493.4	87.3	20.0	185.0	41.0	826.7
Bahamas	-	13.0	-	-	-	13.0
Bolivia	53.5	-	-	-	-	53.5
Brazil	1,524.8	701.6	579.0	256.9	86.6	3,148.9
Cameroon	10.0	-	-	-	-	10.0
Colombia	108.0	170.0	90.0	-	14.3	382.3
Costa Rica	10.0	11.0	-	-	-	21.0
Dominican Republic	20.0	15.0	4.0	-	-	39.0
El Salvador	50.0	-	-	-	-	50.0
Egypt	280.0	-	-	-	-	280.0
Gabon	35.0	89.0	25.0	10.0	-	159.0
Greece	443.0	605.6	270.0	65.0	-	1,383.6
Guinea	-	-	40.0	-	-	40.0
Guyana	15.0	12.5	-	-	-	27.5
Hong Kong	37.1	41.5	10.0	-	-	88.6
India	-	10.0	-	-	-	10.0
Israel	-	-	10.0	-	-	10.0
Ivory Coast	63.0	99.0	-	22.0	-	184.0
Jamaica	135.0	35.6	10.0	-	-	180.6
Kenya	-	-	15.0	-	-	15.0
Korea (South)	147.5	218.0	100.0	49.5	25.0	540.0
Lebanon	40.0	20.0	-	-	-	60.0
Malaysia	140.0	-	50.0	24.0	-	214.0
Malawi	-	5.3	-	-	-	5.3
Mauritania	-	-	-	8.0	-	8.0
Mexico	860.4	1,737.5	196.5	295.0	56.0	3,145.4
New Caledonia	-	-	-	40.0	-	40.0
Nicaragua	21.4	92.0	15.0	10.0	-	138.4
Pakistan	-	-	-	-	17.0	17.0
Panama	77.0	191.0	40.0	16.0	30.0	354.0
Peru	574.5	571.0	138.7	-	-	1,284.2
Philippines	814.0	132.0	50.0	-	5.0	1,001.0
Ras Al-Khaimah	20.0	-	-	-	-	20.0
Senegal	20.0	70.0	-	-	-	90.0
Singapore	24.2	-	-	-	-	24.2
Sudan	231.5	-	-	-	-	231.5
Swaziland	-	-	3.2	-	-	3.2
Taiwan	268.0	-	-	-	-	268.0
Thailand	9.7	-	-	-	-	9.7
Trinidad-Tobago	-	42.0	-	-	-	42.0
Turkey	-	25.3	-	-	-	25.3
Zaire	332.6	225.5	107.0	30.0	25.0	720.1
Zambia	-	150.0	25.0	-	-	175.0
PICA	-	-	-	10.0	-	10.0
ADELA	24.0	60.0	30.0	-	-	114.0
Central American Bank for Economic Cooperation	14.0	-	-	-	-	14.0
Total:	6,896.6	5,430.7	1,828.4	1,021.4	299.9	15,477.0

Table 9.6
Publicly Announced Eurocredits to Developed Countries (in millions of U.S. dollars)

	1974	1973	1972	1971	1970	Total 1970-1974
Australia	127.0	2.8	155.0	239.5	53.1	577.4
Austria	160.0	-	10.0	-	-	170.0
Canada	230.0	101.0	29.0	25.0	45.0	430.0
Denmark	393.5	192.0	41.0	171.5	27.0	825.0
Finland	308.5	406.3	20.0	47.0	36.0	817.8
France	3,155.0	50.0	176.0	40.0	19.5	3,440.5
Germany	10.0	-	-	15.0	20.0	45.0
Iceland	32.5	-	-	-	-	32.5
Ireland	307.0	-	-	-	30.0	337.0
Italy	2,306.5	4,731.5	933.2	317.0	1,387.4	9,675.6
Japan	292.0	160.0	80.0	35.0	5.0	572.0
Netherlands	558.0	250.0	-	-	260.0	1,068.0
New Zealand	490.0	-	31.6	30.0	80.0	631.6
Norway	430.1	194.9	128.0	15.0	-	768.0
Portugal	205.0	10.0	-	-	-	215.0
South Africa	445.8	498.0	149.0	226.1	60.0	1,378.9
Spain	1,124.9	384.9	135.6	317.0	230.0	2,192.4
Sweden	303.5	91.0	115.0	38.0	47.9	595.4
Switzerland	10.0	75.0	-	-	50.0	135.0
United Kingdom	5,367.1	2,781.5	689.0	461.6	48.0	9,347.2
United States	2,174.8	1,572.0	864.5	428.0	1,827.3	6,866.6
EIB	-	-	-	50.0	-	50.0
EKOFISK	230.0	750.0	40.0	80.0	-	1,100.0
Frigg N. Sea Financing	300.0	-	-	-	-	300.0
Unallocated tanker Financing	60.6	-	-	-	-	60.6
Total:	19,021.8	12,250.9	3,596.9	2,535.7	4,226.2	41,631.5

Table 9.7
Publicly Announced Eurocredits to Socialist Countries (in millions of U.S. dollars)

	1974	1973	1972	1971	1970	Total 1970-1974
Bulgaria	160.0	115.0	25.0	-	-	300.0
Cuba	119.7	15.0	23.3	-	-	158.0
East Germany	-	-	35.0	-	-	35.0
Hungary	150.0	90.0	50.0	50.0	30.0	370.0
Korea (North)	-	51.0	-	-	-	51.0
Poland	495.0	270.0	-	-	-	765.0
Rumania	-	-	-	5.0	7.7	12.7
USSR (Comecon)	255.0	50.0	140.0	11.0	-	456.0
Yugoslavia	656.5	243.0	225.6	-	10.0	1,135.1
Total:	1,836.2	834.0	498.9	66.0	47.7	3,282.8

Table 9.8
Publicly Announced Eurocredits to Oil-producing Countries (in millions of U.S. dollars)

	1974	1973	1972	1971	1970	Total 1970-1974
Abu Dhabi	5.1	190.0	-	-	-	195.1
Algeria	150.0	1,563.2	172.0	50.0	-	1,935.2
Bahrain	-	15.0	-	-	-	15.0
Brunei	-	-	27.5	-	-	27.5
Dubai	40.0	120.0	18.3	-	-	178.3
Ecuador	-	8.0	90.0	-	-	98.0
Indonesia	307.5	167.0	92.6	-	-	567.1
Iran	112.5	723.7	335.4	211.0	123.0	1,505.6
Nigeria	20.0	-	-	-	-	20.0
Oman	-	60.0	-	-	-	60.0
Sharjah	86.0	-	-	-	-	86.0
Venezuela	57.5	78.1	200.0	78.9	23.3	437.8
Total:	778.6	2,925.0	935.8	339.9	146.3	5,125.6

Table 9.9.
Publicly Announced Eurocredits, by Country Groups (in millions of U.S. dollars)

	1974	1973	1972	1971	1970	Total 1970-1974
Non-oil exporting less-developed countries	6,896.6 24.2%	5,430.7 25.4%	1,828.4 26.6%	1,021.4 25.7%	299.9 6.4%	15,477.0 23.6%
Oil-exporting less-developed countries	778.6 2.7%	2,925.0 13.6%	935.8 13.7%	339.9 8.6%	146.3 3.1%	5,125.6 7.8%
Developed countries	19,021.8 66.7%	12,250.9 57.1%	3,596.9 52.4%	2,535.7 64.0%	4,226.2 89.5%	41,631.5 63.6%
Socialist countries	1,836.2 6.4%	834.0 3.9%	498.9 7.3%	66.0 1.7%	47.7 1.0%	3,282.8 5.0%
Total:	28,533.2 100.0%	21,440.6 100.0%	6,860.0 100.0%	3,963.0 100.0%	4,720.1 100.0%	65,516.9 100.0%

Source of Table 9.5 through Table 9.9: Morgan Guaranty Trust Co. of New York.

Table 9.10 shows the LDCs' increasing burden of debt service payments, which worries many Eurobankers.

The already low per capita GDP (gross domestic product) of non-oil LDCs is likely to decline in 1975 even with foreign capital inflows of $42.7 billion. According to an estimate made by the World Bank, for the 1976-80 period, the same amount of foreign capital inflows to non-oil LDCs as in 1975 (in real terms) will lead to an average annual increase in per capita GDP of only 1.2

Table 9.10
Debt and Debt Service for 80 Developing Countries, 1967-1972
(millions of U.S. dollars)

	External Public Debt Outstanding, Disbursed and Undisbursed					
	1967	1968	1969	1970	1971	1972
OPEC (8)	7,326.1	8,448.5	10,010.1	12,083.7	14,989.7	17,666.6
Third World (39)	23,446.2	26,318.2	29,424.2	33,835.7	39,039.9	46,579.8
Fourth World (33)	14,687.9	16,644.7	18,052.0	19,795.9	22,129.4	24,101.6
TOTAL (80 Developing Countries)	45,460.2	51,441.4	57,486.3	65,715.3	76,159.0	88,348.0
	Debt Service Payments*					
	1967	1968	1969	1970	1971	1972
OPEC (8)	301.1	399.4	521.9	712.9	1,020.6	1,510.8
Third World (39)	2,347.2	2,786.0	3,130.7	3,644.5	3,836.2	4,393.8
Fourth World (33)	771.2	835.2	957.3	985.1	1,204.6	1,239.4
TOTAL (80 Developing Countries)	3,419.5	4,020.6	4,609.9	5,342.5	6,061.4	7,644.0

*Amortization and interest payments.

Sources: "World Debt Tables: External Public Debt of LDCs," World Bank Doc. EC-167/74, December 15, 1974, pp. 1-4; and Overseas Development Council, *The U.S. and World Development: Agenda for Action, 1975,* (New York: Praeger Publishers, 1975), p. 274.

percent for the low income LDCs (with per capita annual incomes of less than $200) and of 2.8 percent for the middle-income LDCs (with per capita incomes of $200 and above).[5] In order to raise the economic growth rates of non-oil LDCs to the target growth rates of the Second Development Decade, at least for the remaining years of the decade of the 1970s, a substantial increase in capital inflows into developing countries will be required, amounting to an average annual inflow of $61.7 billion, as shown in Case II of the following table (Table 9.11). On the other hand, Case I shows the per capita GDP growth rates assuming no real-term growth in capital inflows.

Table 9.11
Annual GDP Growth Rates, Per Capita, in Non-oil LDCs, and Related Capital Flows (in billions of current dollars)

	Average '69-'73	1974	1975	Average '76-'80 Case I	Average '76-'80 Case II
Growth Rates					
Low-Income Countries[1]	.5%	−.5%	−.7%	1.2%	3.2%
Middle-Income Countries[2]	4.5%	3.9%	−1.2%	2.8%	3.8%
Capital Flows					
Net Official Capital	$7.9	$15.0	$19.8	$24.0	$35.2
Private Capital	7.5	19.4	22.9	25.4	26.5
Total	15.4	34.4	42.7	49.4	61.7

[1]Countries with a per capita income below $200, with a total population of 1 billion.

[2]Countries with a per capita income of $200 and over with a total population of 725 million.

Source: Robert S. McNamara, Address to the Board of Governors at the World Bank Annual Meeting, Washington, D.C., September 1, 1975.

Even to achieve the very low rates of growth envisaged in Case I, non-oil developing countries will have to borrow large sums from private capital markets at the average annual rate of $25.4 billion. Given the worsened prospect for their trade balances, it will be very difficult for non-oil LDCs to borrow the required amounts. It will be even more difficult for these countries to borrow from private markets the necessary amounts envisaged in Case II. This is the reason why most of the non-oil LDCs, especially the poorer ones, will have to depend on concessional aid from both national and international official institutions to meet additional capital requirements. Since the industrial countries have better mechanisms of adjustment to the oil crisis than non-oil LDCs, the former should take urgent measures, in cooperation with the capital-surplus OPEC countries, to "recycle" the necessary capital back to the needy non-oil LDCs.

In 1974, the capital-surplus OPEC countries displayed a great willingness to extend official aids to non-oil LDCs. According to an estimate by the OECD Secretariat, official development assistance actually disbursed by the OPEC members reached $2,603 million in 1974, compared to $11, 300 million from the 17 nations belonging to the OECD's Development Assistance Committee (DAC). In absolute terms, the amount of OPEC development aid is still far less than that of the industrialized nations. In terms of GNP, however, it is much greater than that of the DAC countries; $2,603 million is equivalent to 2.2 percent of the combined GNP of the ten major OPEC donor countries and 3.3 percent of their combined oil revenues during 1974.

By comparison, the foreign aid of $11,300 million from the DAC countries is equivalent to only 0.33 percent of their combined GNP. In fact, the proportion of the industrial nations' GNP devoted to official development assistance has shown a steady decline, from 0.52 percent in 1960 to 0.29 percent in 1975 (see table 9.13 p. 166). Furthermore, the total bilateral aid commitments made by OPEC countries in 1974, including general pledges as well as statements of intention expressed in connection with state visits and similar occasions, amount to $7,172 million, or three and a half times the actual bilateral aid of $2,031 million disbursed in 1974.

When we add to the 1974 bilateral aid disbursements the amount of contributions to the IMF's oil facility, purchases of bonds issued by the World Bank and other regional development banks, and investments and loans to non-oil LDCs on non-concessional

Table 9.12
OPEC Aid Commitments and Disbursements, 1974

	Commitments					Disbursements*				
				As Percentage of:					As Percentage of:	
	Bilateral	Multi-lateral	Total	Oil Revenues	GNP	Bilateral	Multi-lateral	Total	Oil Revenues	GNP
	($ millions)					($ millions)				
Algeria	31	108	139	3.8	1.7	15	45	60	1.1	0.7
Iran	2,802	173	2,975	17.1	10.1	600	2	602	3.4	2.0
Iraq	222	58	280	4.1	2.9	80	35	115	1.7	1.2
Kuwait	957	384	1,341	19.1	15.8	460	70	530	7.6	6.2
Libya	178	241	419	5.5	3.4	45	60	105	1.4	0.9
Nigeria	1	16	17	0.2	0.2	1	15	16	0.2	0.1
Qatar	97	60	157	9.8	7.4	40	20	60	3.7	2.8
Saudi Arabia	2,568	453	3,021	15.1	13.4	650	115	765	3.8	3.4
United Arab Emirates	296	182	478	11.6	10.0	120	45	165	4.0	3.5
Venezuela	20	726	746	7.0	3.6	20	165	185	1.7	0.9
Total	7,172	2,401	9,573	12.1	8.2	2,031	572	2,603	3.3	2.2

*Estimate.

Note: Figures are subject to revision. Some of the disbursement estimates are highly tentative, as are all the comparisons of commitments and disbursements with oil revenues and GNP.

Source: Organization for Economic Co-operation and Development, Development Assistance Directorate, "Flow of Resources from OPEC Members to Developing Countries," Document No. DD-403, December 6, 1974, Table 2, p. 8.

Table 9.13
Net Flow of Official Development Assistance from DAC Countries as a Percentage of Gross National Product

	1960	1965	1970	1971	1972	1973	1974	1975	1976	1977	1978	1979	1980 Case I
Australia	.38	.53	.59	.53	.59	.44	.55	.55	.56	.57			
Austria		.11	.07	.07	.09	.15	.18	.16	.16	.16			
Belgium	.88	.60	.46	.50	.55	.51	.49	.55	.57	.59			
Canada	.19	.19	.42	.42	.47	.43	.50	.51	.56	.59			
Denmark	.09	.13	.38	.43	.45	.48	.54	.57	.61	.64			
Finland		.02	.07	.12	.15	.16	.18	.20	.22	.23			
France	1.38	.76	.66	.66	.67	.58	.60	.60	.61	.61			
Germany	.31	.40	.32	.34	.31	.32	.37	.35	.33	.31			
Italy	.22	.10	.16	.18	.09	.14	.14	.14	.14	.13			
Japan	.24	.27	.23	.23	.21	.25	.25	.23	.22	.21			
Netherlands	.31	.36	.61	.58	.67	.54	.62	.72	.76	.76			
New Zealand[a]			.23	.23	.25	.27	.30	.36	.41	.47			
Norway	.11	.16	.32	.33	.43	.42	.57	.61	.65	.69			
Sweden	.05	.19	.38	.44	.48	.56	.72	.75	.78	.81			
Switzerland	.04	.09	.15	.12	.21	.16	.14	.15	.15	.15			
United Kingdom	.56	.47	.37	.41	.39	.34	.38	.33	.30	.30			
United States[b]	.53	.49	.31	.32	.29	.23	.25	.23	.20	.17			
GRAND TOTALS													
—ODA ($b-Nominal Prices)	4.6	5.9	6.8	7.7	8.5	9.4	11.3	12.2	13.5	15.0	16.6	18.6	20.7
—ODA ($b-Constant 1975 Prices)	10.3	12.3	12.6	13.2	13.4	11.5	12.7	12.2	12.3	12.6	12.9	13.4	14.0
—GNP ($t-Nominal Prices)	.9	1.3	2.0	2.2	2.6	3.1	3.4	3.8	4.4	5.0	5.7	6.4	7.3
—ODA as % GNP	.52	.44	.34	.35	.33	.30	.33	.32	.31	.30	.29	.29	.28
—ODA Deflator	.45	.48	.54	.58	.64	.81	.89	1.00	1.10	1.19	1.29	1.38	1.48

[a]New Zealand became a member of DAC in 1973. ODA figures for New Zealand are not available for 1960-1971.

[b]In 1949, at the beginning of the Marshall Plan, U.S. Official Development Assistance amounted to 2.79% of GNP.

Note: Countries included are members of OECD Development Assistance Committee, accounting for more than 95% of total Official Development Assistance. Figures for 1973 and earlier years are actual data. The projections for 1974 and 1975 are based on World Bank estimates of growth of GNP, on information on budget appropriations for aid, and on aid policy statements made by governments. Because of the relatively long period of time required to translate legislative authorizations first into commitments and later into disbursement, it is possible to project today, with reasonable accuracy, ODA flows (which by definition represent disbursements) through 1975.

Source: Robert S. McNamara, Address to the Board of Governors of the World Bank Group, Washington, D.C., September 30, 1974 (Washington, D.C.: World Bank. 1974).

terms, the total flow of financial resources in 1974 from OPEC countries to non-oil LDCs amounts to $4,750 million, or 3.4 percent of their combined GNP in that year. This compares with $26,350 million in the total financial flows in 1974 from the 17 DAC countries to developing countries, which is equivalent to only 0.77 percent of their combined GNP. In the aggregate, therefore, OPEC's development aid far surpasses the United Nations targets, which require that official development assistance should amount to at least 0.7 percent of a donor country's GNP, and that total financial transfers to developing countries should amount to at least 1 percent of a donor's GNP.

Table 9.14
Official Development Assistance and Total Financial Flows to LDCs (in millions of U.S. dollars)

	1970	1973	1974 Amount	1974 % of GNP	Concessional Element
Official Devel. Assistance					
DAC countries[a]	6,845	9,408	11,300	0.33%	87%[c]
OPEC countries[b]	379	531	2,540	1.8 %	57%[d]
Total Financial Flow					
DAC countries[a]	15,737	24,429	26,350	0.77%	
OPEC countries[b]	650	1,350	4,750	3.4 %	

[a]Development Assistance Committee, comprising 17 OECD industrialized nations.

[b]Ten major aid donors of the OPEC countries

[c]As of 1973.

[d]As of 1974.

Source: OECD, Paris.

As Table 9.15 (p. 168) shows, however, most of OPEC's bilateral aid in 1974 went to Arab countries and four Asian countries, Pakistan, India, Sri Lanka, and Bangladesh. These countries received 95 percent of OPEC's total bilateral aid disbursed in 1974 and 96 percent of OPEC's total aid commitments made in 1974. It is unlikely that OPEC countries can maintain their development aid at a level comparable to that in 1974 throughout

Table 9.15
Recipients of Bilateral Aid from OPEC Countries, 1974

	Commitments		Disbursements*	
	($ millions)	(as percentage of total)	($ millions)	(as percentage of total)
Arab Countries				
Egypt	3,121*	43.5*	765	37.7
Syria	1,003*	14.0*	325	16.0
Jordan	185*	2.6*	140	6.9
Mauritania	153*	2.1*	25	1.2
Sudan	107	1.5	70	3.4
Somalia	82	1.1	25	1.2
Morocco	80	1.1	25	1.2
Tunisia	54	0.8	15	0.7
Bahrain	21	0.3	10	0.5
Yemen, Arab Rep.	19	0.3	15	0.7
Yemen, People's Dem. Rep.	12	0.2	10	0.5
Other	8	0.1	8	0.4
Africa				
Malagasy Republic	114*	1.6*	–	–
Guinea	16	0.2	5	0.2
Uganda	12	0.2	2	0.1
Senegal	11	0.1	5	0.2
Other	79	1.1	46	2.3
Asia				
Pakistan	957*	13.3*	355	17.5
India	945	13.2	75	3.7
Sri Lanka	86	1.2	35	1.7
Bangladesh	82	1.1	50	2.5
Latin America				
Guyana	15	0.2	15	0.7
Honduras	5	0.1	5	0.2
Europe				
Malta	5	0.1	5	0.2
Total	7,172*	100.0	2,031	100.0

*Estimate.

Note: Figures are subject to revision. Breakdown between Egypt, Syria, and Jordan is tentative. Commitment figures for Pakistan include $200 million by Saudi Arabia and Kuwait which were not made public but are believed to have been made.

Source: Organization for Economic Co-operation and Development, Development Assistance Directorate, "Flow of resources from OPEC members to Developing Countries," Document No. DD-403, December 6, 1974, Table 4, p. 10.

the remainder of this decade. The OPEC members themselves are developing countries, and the source of their capacity to give foreign aid is not their combined GNP, which is only a small fraction of that of the OECD countries, nor even their GNP per capita, which on average is also substantially less than that of the OECD members, but rather their large surplus trade balances due to sharp oil-price increases. By 1980, however, only a few of the OPEC countries are expected to have substantial current-account surpluses. Consequently, official aid from the OPEC countries is likely to diminish in the coming years along with their declining oil money surpluses.

In the meantime, most non-oil LDCs are caught up in a critical dilemma due to the oil-price increases. Historically, energy consumption in the LDCs has increased in close parallel with their economic growth. Studies of the relationship between economic growth and energy demand in the LDCs suggest that a 10 percent increase in GNP will produce an increase in energy demand of 13 to 16 percent.[6] This means that non-oil LDCs will be hard-pressed to continue their economic development in the face of sharply higher oil prices than a few years ago. Their plight has been intensified not only by the sudden upsurge in the cost of oil, but also by the continuing worldwide inflation, the deterioration of their terms of trade, and the prolonged recession in their export markets. The net effect of these external forces has been a drastic decline in their prospective economic growth rates and an unbearable burden due to their increased foreign-exchange requirements. In this situation, it is up to the industrial countries, as well as the surplus OPEC countries, to take decisive measures in the spirit of enlightened self-interest to stave off the impending economic collapse of the non-oil developing countries.[7]

NOTES

[1] *World Financial Markets,* Morgan Guaranty Trust Co. of New York, October 20, 1975, p. 5; and *The Economist,* August 16, 1975, p. 60.

[2]The following classification of non-oil LDCs draws heavily on an excellent paper by James P. Grant, "Energy Shock and the Development Prospect," in James W. Howe and the staff of the Overseas Development Council, *The U.S. and the Developing World: Agenda for Action 1974,* (New York: Praeger Publishers, 1974), pp. 31-50.

[3]Richard O'Brien, "Assessing the Credit Risks of the Developing Countries," *Euromoney,* October 1975.

[4]Pierre Latour, "Euromarkets Wait for LDCs' Credits to be Repaid," *Euromoney,* October 1975.

[5]Robert S. McNamara, Address to the Board of Governors at the World Bank Annual Meeting, Washington, D.C., September 1, 1975.

[6]Sam Schurr, ed., *Energy, Economic Growth, and the Environment,* Resources for the Future (Baltimore: Johns Hopkins University Press, 1972), pp. 180-83.

[7]For some of the policy proposals dealing with the non-oil developing countries, see Robert M. Dunn, Jr., "The Less Developed Countries," in Joseph Yager and Eleanor Steinberg, eds., *Energy and U.S. Foreign Policy* (Cambridge, Mass.: Ballinger Publishing Co., 1974), pp. 163-81; and Constantine Michalopoulos, "Financing Needs of Developing countries: Proposals for International Action," *Essays in International Finance,* No. 110, Princeton University, June 1975.

Chapter 10
Conclusion

This study has analyzed the effects of the quadrupling of the price of oil on the international economy. At the same time, we have attempted to develop a set of policies to meet the economic challenge posed by the sharp oil-price upsurge. Oil is such an essential raw material for the world economy that the sudden and sharp increase in its price has had significant effects on both oil-exporting and oil-importing countries. The full impact of the oil-price increase may take years to unfold and thus may be difficult to gauge precisely at this time. However, what this study has attempted to do is find answers to such important questions as: How has the world oil industry changed over the years to tip the balance finally in favor of the oil-producing countries at the expense of the oil-consuming countries as well as the oil companies? What was the motive underlying the creation of OPEC? How has the oil-price increase affected the real income and balance of payments of the oil-importing countries? What has been the price-increase effect on international liquidity and the Eurocurrency market? Last but not least, how have the non-oil LDCs fared in the aftermath of the oil-price upsurge?

In its attempt to answer the above questions, this study has utilized the theoretic-analytical tools of modern economics and put less reliance upon statistical inferences. This methodological choice has been partly determined by the fact that the current oil crisis developed only a couple of year ago; consequently, we have not enough serial data to run a statistical test of the various economic variables changed by the oil-price increase. Moreover, our main concern in this study has been to set up a conceptual model to explain the *change* in economic parameters, rather than trying to calculate the parameters themselves using an incomplete data base. In the following sections, we will summarize the principal findings of our study and suggest areas for further research in view of these findings.

FINDINGS OF THE STUDY

In the early 1970s, OPEC moved from an essentially defensive role to an offensive one. When OPEC was first established in 1960, the main purpose of the member countries was to prevent a unilateral reduction of crude-oil prices by Western oil companies. During the first ten years of its existence, OPEC maintained this defensive orientation, directing its efforts toward preventing any further erosion in the posted prices of oil. Consequently, no further reduction of the posted prices by oil companies took place after 1960, even though the market prices of oil continued to erode during the 1960s. Until 1970, therefore, the world oil industry had been operating on the basis of a *modus vivendi* between oil companies and oil-exporting countries. Although the posted prices were not reduced by the oil companies due to the united front of the OPEC countries, the market prices of oil were largely determined by the factors of supply and demand, and only the oil companies had direct control over the amount of oil to be produced and supplied to the market. As long as the OPEC members did not possess the power to control and regulate the production volume of oil in their countries, they could not "dictate" oil prices unilaterally.

During the early 1970s, however, the international oil industry underwent revolutionary political and economic changes. The seeming equilibrium of the international oil market was first upset in 1973 by Libya, which took advantage of the increasing importance of Libyan oil to Europe due to a combination of factors such as the closing of the Suez Canal during the Six-Day Arab-Israeli War of 1967, the subsequent increase in world tanker rates, and the accidental shutdown in 1970 of the Trans-Arabian Pipeline which carried Saudi crude oil to the Mediterranean. Encouraged by the Libyan success in forcing Western oil companies to effect a 13-percent raise in the posted price of oil, the OPEC members decided to adopt an offensive negotiating tactic against Western oil companies, and, by 1973, had taken over both crude-oil pricing and production decisions, which had traditionally been under the control of the oil companies. Moreover, many OPEC countries have progressively nationalized their oil industries or acquired majority ownership of oil reserves and producing facilities located within their boundaries.

Since oil imports are highly price-inelastic in the short run, the unilateral price increase of about 400 percent by the OPEC members within one year has led to a sharp intensification of the payments burden of most oil-importing countries. Whether the current high oil prices can be maintained in the future will depend on the solidarity of the OPEC members in defending their price policy as well as on the development of oil substitutes and non-OPEC sources of oil supply. Nevertheless, the OPEC members as a group are expected to accumulate a huge financial surplus. Estimates of cumulative OPEC surpluses by 1980 range from $180 billion to $450 billion in current U.S. dollars.

The oil crisis has caused the severest disturbance in world trade balances since the Second World War, and the deficits of may oil-importing countries pose one the most difficult balance-of-payments adjustment problems in the international monetary system. The huge trade deficits of many oil-importing countries and the even greater trade surpluses of several OPEC countries present policy issues fundamentally different from those normally associated with the problems of international trade imbalances. The internal economic balance of oil-importing countries has been subjected to both deflationary and inflationary biases by the oil-price upsurge: the former due to the "excise tax" effect of the oil-price increase on aggregate consumption, and the latter to the sharp price increase of oil products and oil substitutes as well as the "ratchet effect" on prices of currency depreciation in oil-importing countries.

At the same time, the external balance of oil-importing countries has been disturbed by massive oil payments. But the conventional remedies for an individual oil-importing country, such as currency devaluation or deflationary domestic economic policies, may not work for the oil-importing countries as a group due to the limited absorptive capacity of the oil-exporting countries and the fact that oil payments are made in the currencies of certain oil-importing countries rather than in the currencies of OPEC countries. Consequently, the world economic order is faced with a new dilemma: the internal balances of oil-importing countries may not be restored by macroeconomic policies concerned with the aggregate demand, while their external balances cannot be achieved through conventional adjustment processes.

In seeking to resolve this dilemma, we have developed a new adjustment model that incorporates the time element in order to balance internal economic growth against external balance. In other words, our new model considers adjustment toward internal, as well as external, balance as a dynamic process involving a trade-off between internal economic growth and external equilibrium. Our model demonstrates that the faster the economies within the OPEC group grow in relation to the economies of non-OPEC countries, or the more the income elasticity of demand for oil imports from OPEC countries is lowered, the sooner will the oil deficits of non-OPEC countries disappear.

For the world economy as a whole, our model suggests two kinds of adjustment policies in order to ameliorate or overcome the current oil-induced payments imbalances: (1) to raise the economic growth rate of OPEC countries relative to the growth rate of non-OPEC economies; and (2) to economize oil consumption and/or develop oil substitutes in order to lower the income elasticity of oil imports. It should be noted, however, that our model does not necessarily recommend deflationary policies in non-OPEC countries; it simply requires that the differential in the economic growth rates between OPEC and non-OPEC countries be increased as much as possible. It has been shown that a one-percent increase in the growth-rate of OPEC economies will shorten the required adjustment period for oil deficits by 20 percent, given the growth rate of non-OPEC economies.

The sharp increase in oil prices has also had a negative effect on the real incomes of oil-importing countries due to three factors. First, the terms of trade have worsened for oil-importing countries, with an attendant transfer of real income from oil-importing countries to oil-exporting countries. Second, the increase in oil prices has had an excise-tax effect on consumption in oil-importing countries, thus depressing real income through a reduction in the income multiplier. Third, the worsened trade balances of the oil-importing countries tend to encourage the governments of these countries to adopt deflationary policies to reduce their payments deficits, thereby inducing slower economic growth than would be the case without the oil crisis.

Apart from these three negative effects on the absolute level of real incomes in the oil-importing countries, there is also a distributional effect on real incomes among the citizens of the oil-consuming countries due to the increased cost of energy. Furthermore, because of the more or less simultaneous appearance of the above negative real-income factors in all the oil-importing countries, cumulative international processes have emerged which reinforce the worldwide deflationary trend. The rise in oil prices simultaneously exerts deflationary influences on all oil-importing countries, and such a worldwide synchronization of deflationary impact leads to an amplification of its effects through international economic linkages. Thus, we have to consider the deflationary effect on a country from the international multiplier as well as from the respective domestic income multiplier. This indicates that the current oil crisis is a truly global problem, which requires global, rather than individual-national measures to find an effective solution.

Increased oil-money flows as a result of the price hikes have contributed to an increase in the size of the Eurocurrency market. In 1974, for example, OPEC countries supplied $23 billion, and oil companies several billion dollars, directly to the Eurocurrency market. However, the growth of the Eurocurrency market due to oil-money inflows may have been higher than $25 billion because of the Eurocurrency multiplier effect. On the demand side, most of the explosive growth of the floating-rate Eurocredit volume in the first half of 1974 was accounted for by precautionary borrowings on the part of oil-importing countries to accumulate external funds for future financing of oil deficits.

International reserves increased by 18 percent in 1974, comparable to the growth rate recorded in 1972 (19 percent) and substantially higher than that in 1973 (4 percent). The reason why total international reserves expanded rapidly in 1974 was that the process of international reserve creation took place when the reserves transferred to the oil-exporting countries were replenished by oil-importing countries through borrowing in the United States or the Eurocurrency market. In this connection, both the United States and the Eurocurrency market have enlarged their role as the principal reserve suppliers to the world. However, almost the entire reserve increase recorded in 1974 accrued to OPEC countries, while the reserves of seventeen industrial countries as a group showed an increase of only $2.6 billion, and those of non-oil LDCs hardly increased at all.

The oil-price upsurge has also contributed to world inflation. First, there is the direct effect on inflation of price increases in both oil and oil products. Second, there is the effect of the price increase of oil substitutes. Third, there is the "ratchet effect" of the currency depreciation of oil-importing countries. And fourth, wages have had to keep pace with the rising inflation rate caused by the oil-price increase. On the other hand, however, oil-price increases have had some deflationary effect, analogous to that of an excise tax, but this effect has not been strong enough to offset the overall inflationary bias of the oil-price upsurge.

Perhaps the most serious economic effects of the sharp increase in oil prices have been felt by those countries that are least able to afford such an increase, that is, the non-oil developing countries. The abrupt rise in oil prices directly added about $8.5 billion to the import bills of non-oil LDCs in 1974, and, consequently, they suffered a combined trade deficit of $26 billion for the year. Because of the continuing recession in the Western industrial countries, non-oil LDCs are expected to suffer an even worse trade deficit, amounting to between $34 and $42 billion, in 1975. In addition, their ability to borrow in private capital markets is severely limited due to their deteriorating credit ratings.

In order to ameliorate somewhat the plight of non-oil LDCs, various recycling proposals have been advanced. Only one, the IMF Oil Facility, has been operational so far; however, even this is to cease operation by 1976. Therefore, if non-oil LDCs are to be given any reasonable chance to "muddle through" the current oil-caused payments crisis,[1] official capital flows from both industrial and OPEC countries to non-oil LDCs have to be increased. In the absence of such capital flows, non-oil developing countries may be forced to abandon any hope of achieving even a modicum of economic growth in the coming decade.

AREAS FOR FURTHER RESEARCH

Obviously, the study of a subject as complicated as the implications of recent oil-price hikes for the world economy can be completed neither in one volume nor by one economist. The subject contains too many unknowns and too much new ground for a single researcher to cover adequately; therefore, this study should be considered only as introduction. At least, an attempt has been made here to identify all the major economic issues connected with the oil crisis. Further investigation of these issues is needed to understand the full implications of the oil-price upsurge for the world economy.

In this connection, we may suggest some of the specific problem areas where more research is especially needed. First, the future political economy of oil is an important topic, as oil economics increasingly takes on the shape of a bilateral monopoly between the governments of oil-exporting and oil-importing countries.

Western oil companies are being relegated to the status of mere agents between the producing and consuming countries, while the strategic decisions on oil are being made more and more at the government level. Since the oil-exporting countreis have formed a monopoly of sellers, we may soon witness the emergence of a Galbraithian countervailing force in the form of a monopsony of oil-importing countries allied against OPEC. The future development of oil prices will then critically depend on the way both the oil-producing and the oil-consuming countries maneuver against each other in a classic game of bilateral monopoly.

Second, there is a need for more rigorous testing of the various hypotheses presented by our dynamic adjustment model, described in Chapter 4. We have been hampered in this attempt by a lack of reliable data for the parameters, such as the income elasticities of demand for imports in both OPEC and non-OPEC countries. Through up-to-date econometric calculation of these parameters, we may be able to arrive at a more precise estimate of the required adjustment period for eliminating oil deficits. By calculating these parameters for each country and the world as a whole, the generalized dynamic adjustment model can also be put to use in evaluating the adequacy or inadequacy of the international reserves of individual countries.

Finally, more work needs to be done to find practical ways for non-oil LDCs to cope with the oil-caused payments crisis. It would be a great mistake to conclude, by just looking at the global economic aggregates, that the world economy will somehow be able to muddle through the current crisis. For such a view neglects the weakest link in the chain of international economic interdependence. In the general tendency to look only at the major economic issues raised by the oil-price upsurge, economists are likely to overlook the critical economic realities faced by the most seriously affected of the non-oil LDCs. However, it is this group of countries, rather than the OECD countries, that is likely to spark an economic disaster of international scale, if there is to be one. In general, economists tend to leave this urgent practical problem to the politicians, who are primarily preoccupied with the economic problems of their own countries. What is required of economists now, therefore, is a new bold vision comparable to that of the framers of the Bretton Woods system during the turmoils of the Second World War.

NOTES

[1]Gerald A. Pollack, "Are the Oil Payments Deficits Manageable?" *Essays in International Finance,* No. 111, Princeton University, June 1975, pp. 23-24.

Appendix A

ARAB FUNDS AT THE END OF 1974[1]

1. Abu Dhabi Fund for Arab Economic Development (ADFAED)

Capital: subscribed - $506 million; paid up - $125 million
Headquarters: Abu Dhabi
Shareholders: Emirate of Abu Dhabi
Purpose: Financing of development projects in developing Arab countries
Publicized loans in 1974: $30.5 million

2. Arab Fund for Economic and Social Development (AFESD)

Capital: fully paid - $350 million
Headquarters: Kuwait
Shareholders: 20 Arab countries
Purpose: (a) Providing facilities for economic and social development projects, mainly in Arab developing countries
(b) Financing of projects to exploit mineral resources in Arab countries
Publicized loans in 1974: $166 million

3. Arab Bank for African Economic Development

Capital: $231 million
Headquarters: Khartoum
Shareholders: main Arab oil exporting countries
Purpose: Financing of development projects in developing African countries
Publicized financial aid in 1974: $13 million

4. Arab Fund for Technical Assistance to Africa

Capital: $25 million
Shareholders: Arab League

5. Kuwait Fund for Arab Economic Development (KFAED)

Capital: subscribed - Kuwaiti Dinars 1,000 million ($3.5 billion)
paid up - Kuwaiti Dinars 870 million ($3 billion)

[1]International Chamber of Commerce, *Report on the Financial Implications of the Oil Crisis,* Paris, February 18, 1975.

Headquarters: Kuwait
Shareholder: Emirate of Kuwait
Purpose: Providing loans to Arab countries and developing countries in Africa and Asia
Publicized loans in 1974: $190 million

6. Iraq Fund for External Development (IFED)

Capital: $170 million
Heaquarters: Baghdad
Shareholder: State of Iraq
Purpose: Financing of projects in Arab and other developing countries

7. Islamic Development Bank

Capital: subscribed - $2,000 million
Capital: paid up - $900 million
Headquarters: Riyad
Shareholders (23): Algeria, Bangladesh, Chad, Egypt, Indonesia, Jordan, Kuwait, Lebanon, Malaysia, Mali, Mauritania, Morocco, Oman, Pakistan, Qatar, Saudi Arabia, Senegal, Somalia, Sudan, Tunisia, Turkey, UAE, North Yemen.
Purpose: Providing interest free loans to member states for economic and social development projects.

8. Saudi Arabian Development Fund

Capital: $2,800 million (intitial capital)
Headquarters: Riyad
Shareholders: Kingdom of Saudi Arabia
Purpose: Providing financial aid for development projects in Arab and non-Arab developing countries.
Publicized loans in 1974: $161 million

9. Arab Investment Company

Capital: $200 million
Headquarters: Riyad
Branches: Cairo, Khartoum
Shareholders: Saudi Arabia, Abu Dhabi, Algeria, Bahrain, Egypt, Kuwait, Qatar, Sudan, Syria.
Purpose: Financing of development projects in the Arab world.

10. Arab-African Oil Assistance Fund

Capital: subscribed - $200 million
paid up - $130 million

Contributors: Main oil exporting Arab countries, i.e., Kuwait with $30 million, Saudi Arabia with $40 million, and UAE with $20 million.

Purpose: This fund is managed by the African Development Bank in Abidjan. It provides financial aid for African countries to help them overcome trade deficits caused by higher oil prices.

Publicized credits in 1974: $81.5 million

Appendix B:

JOINT MIDDLE EAST/OECD BANKING VENTURES*

Joint ventures by the financial institutions of OECD and Middle East countries have increased in importance in recent years. The number of these joint ventures more than doubled between end-1973 and August 1975, and their participation in international business tended to grow correspondingly. Their range of activity has become wider, encompassing a growing number of markets and services. In particular, in addition to the traditional activity of taking and rechannelling funds from Middle East investors, joint ventures have become increasingly interested in merchant banking and the syndication of international bond issues and medium-term Euro-loans, as well as in portfolio management and providing advisory and brokerage services. Although the majority of ventures are still located in the Middle East (see table below), the number of institutions established in European financial centers—especially London (5), Luxembourg (5), and Paris (3)—has tended to grow rapidly. Most recently, Middle East institutions have shown an interest in creating links with the Far East, and three companies have been set up in Hong Kong.

Identified Joint Ventures (End of Period)

	1960	1970	1973	Aug. 1975
Number of banks (incl. affiliates)	9	14	24	49
Ventures located in the Middle East	9	12	14	29

The Secretariat has identified forty-nine institutions bringing together Middle East and OECD partners. The main characteristics of existing joint ventures are given below. For each institution the following information is provided:

- official name of the institution
- date of establishment (Est.)

* *Financial Market Trends,* OECD, Paris, September 1975.

- location
- capital (C)
- major shareholders**

To help readers to locate banks, the descriptive list is preceded by an index.

INDEX

**Numbers in brackets after the name of shareholders show percentage holdings at the end of August 1975.

(A) EUROPEAN - MIDDLE EASTERN

1. Bank of Tehran

Est.: 1952; Tehran; C: R 800m;
Shareholders: Iranian public (65); Banque de Paris et des Pays-Bas (15).

2. Foreign Trade Bank of Iran

Est.: 1958; Tehran; C: R 700m;
Shareholders: Bank Melli Iran (20); Banca Commerciale Italiana; Deutsche Bank.

3. Credit Bank of Iran

Est.: 1958; Tehran; C: R 1,000m;
Shareholders: Iranian shareholders (60); Crédit Lyonnais (10); Banque Nationale de Paris (10).

4. Irano-British Bank

Est.: 1958; Tehran; C: R 400m;
Shareholders: Iranian public (60); Chartered Bank -UK- (40).

5. Bank of Iran and the Middle East

Est.: 1959; Tehran; C: R 400m;

Shareholders: Iranian public (60); British Bank of the Middle East -UK- (40).

6. Mercantile Bank of Iran and Holland

Est.: 1959; Tehran; C: R 500m;
Shareholders: Iranian public (65); Algemene Bank Nederland -Netherlands- (35).

7. Union de Banques Arabes et Françaises (UBAF)

Est.: 1970; Paris; C: FF 110m
Shareholders: Crédit Lyonnais (31.95) Banque Française du Commerce Extérieur (8); Arab Bank -Jordan- (6.33); Banque Extérieure d'Algérie (6.33); Commercial Bank of Syria (6.33); Libyan Arab Foreign Bank (6.33); Rafidain Bank -Iraq- (6.33); Central Bank of Egypt (6.33).

UBAF's Affiliates:

(i) UBAF Ltd.

Est.: 1972; London; C: 5m;
Shareholders: Midland Bank -UK- (25); Libyan Arab Foreign Bank (25); UBAF (50).

(ii) Unione di Banche Arabe ed Europee Spa

Est.: 1972; Rome; C: Lit 5,000m;
Shareholders: UBAF (51); Banco di Roma (8.50); Banco Nazionale del Lavoro (8.50); Libyan Arab Foreign Bank (7).

(iii) Union de Banques Arabes et Européennes
Est.: 1973; Luxembourg and Frankfurt; C: DM 30m;
Shareholders: UBAF (33.33); Arab Bank -Jordan-; Arab Bank Overseas Ltd. -Switzerland-; Bayerische Vereinsbank; Commerzbank; Westdeutsche Landesbank Girozentrale.

(iv) Union de Banques Arabes et Nippon

Est.: 1974; Hong Kong; C: HK$ 25m;
Shareholders: UBAF (20); Bank of Tokyo (8); Long Term Credit Bank of Japan (8); Mitsui Bank (8); Nomura Securities (8); Sanwa Bank (8); Alahli Bank of Kuwait (7);

Arab African Bank (7); Arab Bank Ltd. -Jordan- (7); Libyan Arab Foreign Bank (7).

(v) UBAF Financial Services Ltd.

Est.: 1974; London; C: £ 1m;
Shareholders: UBAF (25); Hambros Bank (20); Midland and International Banks (20); Crédit Lyonnais (10); Industrial Bank of Kuwait (10).

(vi) UBAF Arab American Bank

Est.: 1975; New York; C: US$ 25m;
Shareholders: Bankers Trust New Tork Corp. (5); First Chicago Corp. (5); Security Pacific Corp., Los Angeles (5); Texas Commerce Bankshares Inc. (5); Arab and European Shareholdings.

8. Arab and Morgan Grenfell Finance Company

Est.: 1974; London; C: £ 1m;
Shareholders: Morgan Grenfell Holdings -UK- (50); Arab Bank - -Jordan- (45)
Engaged in arranging and financing projects mainly in the Middle East.

9. Oryx Investments Ltd.

Est.: 1974; Dubai; C: US$ 1.25m;
Shareholders: Arab shareholders (60); Arbuthnot Latham -UK- (20); Standard and Chartered Banking Group -UK- (20)
Offers a full range of merchant banking services.

10. Cairo Barclays International Bank

Est.: 1974; Cairo; C: US$ 10m;
Shareholders: Barclays Bank International (50); Banque du Caire (50)
Provides merchant and wholesale banking services, including the financing of foreign trade in support of development projects in Egypt and elsewhere in the Middle East.

11. J. Henry Schroder & Co.

Est.: 1974; Beirut;

Shareholders: J. Henry Schroder Wagg -UK-; Asseily Family -Lebanon-.

12. Société Financière pour le Moyen Orient (SOFIMO)

Est.: 1975; Beirut; C: Lib£ 6m;
Shareholders: Société Générale (42.50); FRAB Bank -France-; Arab Investors.

13. Misr International Bank

Est.: 1975; Cairo; C: US$ 2.5m;
Shareholders: Public subscription by Egyptian nationals (21); Bank Misr -Egypt- (20); Banco di Roma (13); UBAF Ltd. -UK- (10).

14. Compagnie D'Investissement Irano-Française

Est.: 1975; Tehran; C: R 1000m;
Shareholders: Société Générale (40); Industrial and Mining Development Bank of Iran (25); Industrial Credit Bank -Iran- (20); Agricultural Development Bank of Iran (15)
Promotes industrial and agricultural projects in Iran for Iranian and French investors, and explores the possibilities in France for Iranian investments.

15. Banco Arabe Espanol S.A.

Est.: 1975; Madrid; C: Pts 1,500m;
Shareholders: Kuwait Foreign Trading Co. (30); Libyan Arab Foreign Bank (30); Instituto de Crédito Oficial (9.30); Other Spanish concerns (30.40)
Promotes reciprocal flows of investments between Spain and the Arab countries.

(B) EUROPEAN - U.S. - MIDDLE EASTERN

16. National Bank of Dubai

Est.: 1963; Dubai; C: D 18.3m;
Shareholders: Local Dubai merchants (56); National Bank of Kuwait (30); Grindlays Bank -UK- (10); Bank of America (4).

17. Commercial Bank of Dubai

Est.: 1969; Dubai; C: D 20m;

Shareholders: Commerzbank; Chase Manhattan Bank; Commercial Bank of Kuwait.

18. Dubai Bank Ltd.

Est.: 1970; Dubai; C: D 11.5m;
Shareholders: Local investors (75); Wells Fargo -US- (10); Hambros Bank -UK- (10).

19. Bank of credit and Commerce International S.A.

Est.: 1972; Luxembourg; C: US$ 15m;
Shareholders: Geoffry Robert Edward Wallis -UK- (35); Sheikh Zaid bin Sultan al Nahiyan -Abu Dhabi- (20); Bank of America (20); Stock Investment Holding Co. -US- (18).

(i) BCCI Finance International Ltd.

Est.: 1974; Hong Kong; C: HK$ 5m;
To promote the flow of European and Middle Eastern investments towards Hong Kong.

20. International Financial Advisers

Est.: 1974; Kuwait; C: KD 500,000;
Shareholders: Private Kuwaiti individuals (55); Robert Fleming & Co. -UK- (15); William Kent & Co. -USA- (15); Banque Worms -France- (15)
To provide technical financial services to Arab governments and companies, to offer advice to foreign and local borrowers and lenders, and to assist foreign companies wishing to establish operations in the Arab countries.

21. Arab International Finance Company (ARINFI)

Est.: 1974; Switzerland; C: SF 14m;
Shareholders: International Marine Banking Co. -UK- (12.50); Marine Midland Bank -US- (12.50); A.K. Holdings -Luxembourg- Kuwait International Investment Co.
Engaged in investment management, foreign exchange dealings and general international financial operations.

22. Tehran Investment and Finance Co. (TIFIC)

Est.: 1975; Tehran; C: US$ 5m;

Shareholders: Bank of Teheran and other Iranian banks (60); Bayerische Vereinsbank -Germany- (8); Bank of America.

(C) EUROPEAN - JAPANESE - MIDDLE EASTERN

23. Banque Franco-Arabe D'Investissements Internationaux (FRAB)

Est.: 1969; Paris; C: FF 70m;
Shareholders: Kuwait Group (26.50); Société Générale (25); Saudi Arabian Group (12); Société Générale de Banque -Belgium- (7).

(i) FRAB Holdings S.A.

Est.: 1974; Luxembourg; C: LuxF 500m;
Shareholders: as FRAB (Paris).

24. European Arab Holding S.A.

Est.: 1972; Luxembourg; C : LuxF 1,000m;
Shareholders: FRAB Bank (10); Creditanstalt - Bankverein -Austria- (5.69); Deutsche Bank (5.69); Midland Bank (5.69); Banca Commerciale Italiana (5.69); Amsterdam-Rotterdam Bank (5.69); Middle Eastern Countries (39.60.)

Subsidiaries:

(i)European Arab Bank S.A.

Est.: . . . .; Brussels; C: BF 500m;
Shareholders: as European Arab Holding S.A.

(ii)Europaisch-Arabische Bank GmbH

Est.: . . . .; Frankfurt; C: DM 15m;
Shareholders: as European Arab Holding S.A.

25. Banque D'Investissement et de Financement (INFS)

Est.: 1974; Beirut; C: L£ 15m;
Shareholders: Banque Audi -Lebanon- (35); Non-institutional Arab Shareholders (25); Caisse Centrale de Banques Populaires (8); Hambros Bank -UK- (8); Mitsui Bank -Japan- (8); Nomura Securities -Japan- (8)
Provides medium and long-term financing for development projects in the Middle East.

(D) U.S. - MIDDLE EASTERN

26. Iranians' Bank

Est.: 1959; Tehran; C: R 1,000m;
Shareholders: Iranian public (65); First National City Bank (35).

27. Egyptian American Bank

Est.: 1975; Cairo; C: US$ 30m;
Shareholders: Bank of Alexandria -Egypt- (51); American Express (49)
Offers a full range of commercial banking services, including Egyptian-pound deposits. Services primarily oriented towards Egyptian customers and projects.

28. Chase National Bank of Egypt

Est.: 1975; Cairo; C: E£ 10m;
Shareholders: National Bank of Egypt (51); Chase Manhattan Bank (49).

29. Iran Chase International Bank

Est.: 1975; Tehran; C: US$ 50m;
Shareholders: Chase Manhattan Bank (35); Industrial Credit Bank of Iran (35); Iranian public (30)
Assists in the development of Iran and promotes the growth of Tehran as a financial center.

(E) JAPANESE - MIDDLE EASTERN

30. International Bank of Iran and Japan

Est.: 1959; Tehran; C: R 1,000m;
Associated with Bank of Tokyo (Japan).

(F) U.S. - JAPANESE - MIDDLE EASTERN

31. Bank Dariush

Est.: 1974; Tehran; C: R 200m;
Associated with Continental Bank (U.S.A.) and Sanwa Bank (Japan).

(G) EUROPEAN - U.S. - JAPANESE - MIDDLE EASTERN

32. Iran Overseas Investment Bank

Est.: 1973; London; C: £ 8m;
Shareholders: Bank Melli -Iran- (25); Industrial and Mining Development Bank of Iran (25); Bank of America (6.25); Midland Bank (6.25); Manufacturers Hanover Trust (6.25); Bank of Tokyo (6.25); Barclays Bank International (6.25); Industrial Bank of Japan (6.25); Deutsche Bank (6.25); Société Générale (6.25).

33. Development and Investment Bank of Iran

Est.: 1973; Tehran; C: R 2.100m;
Shareholders: Iranian shareholders (80.96); Williams & Glyn's Bank -U.K.- (4.76); Long-term Credit Bank of Japan (4.76).

34. Arab Finance Corporation

Est.: 1973; Beirut; C: L£ 6m;
Shareholders: Manufacturers Hanover Trust (18); Kuwait Investment Co. (18); Investment Promotion Group -Lebanon- (18); Banque de l'Union Européenne -France- (18); Bank of Tokyo (10); Beirut-Riyad Bank -Lebanon- (9);
Developing Arab financial centers and supporting financial projects in the Middle East.

35. Arab Financial Consultants Co.

Est.: 1974; Kuwait; C: KD 500,000;
Shareholders: Kuwaiti Interests (51); Other Arab Investors (25); Arbuthnot Latham -U.K.- (6); Tanjo Kobe Bank (6); Philadelphia National Bank -U.S.A.- (6); Banque de Suez -France- (6);
A financial intermediary channelling Kuwait funds into public issues and private placements for overseas borrowers, and advising on joint ventures.

36. Al-Bank Al-Saudi Al-Alami Ltd.

Est.: 1975; London; C: £ 12.5m;
Shareholders: Saudi Arabian Monetary Agency (SAMA) (50); Morgan Guaranty Trust Co. (20); The Bank of Tokyo Ltd. (5); Banque Nationale de Paris (5); Deutsche Bank A.G. (5); National Westminster Bank Ltd. (5); Union Bank of Switzerland (5);

The bank operates on a purely commercial basis and not as an agent of SAMA, but it may perform some services, including the financing of trade, for the government of Saudi Arabia.

37. Union Bank S.A.L.

Est.: 1975; Beirut; C: L£ 15m;
Shareholders: Gefinor Group -Lebanon- (21.7); Banque de Crédit Agricole, Industriel et Foncier -Lebanon- (13.30); Kredietbank S.A. Luxembourgeoise (20.50); Martin Currie & Co. -U.K.- (15);
Developing money and capital markets in Lebanon and the Middle East, by underwriting and placing government and corporate securities as well as promoting services such as leasing, factoring, money management and advice on mergers and acquisitions.

(H) OTHERS

38. Industrial and Mining Development Bank of Iran

Est.: 1959; Tehran; C: R 3,150m;
Shareholders: Iranian public (82); 22 Foreign Banks (18).

39. Compagnie Arabe et Internationale D'Investissement (CAII)
Est.: 1973; Luxembourg; C: US$ 30m;
Shareholders: Société Financiére Européenne (25); Kuwait Investment Co.; Government of Abu Dhabi; Qatar National Bank; National Commercial Bank -Saudi Arabia-; National Investment Company -Libya-; Banque Nationale de Paris; Banco Central -Spain-; Canadian Imperial Bank of Commerce.

Subsidiary of CAII:
Banque Arabe et Internationale D'Investissement (BAII)
Est.: 1973; Paris; C: FF 50m;
Shareholders: Société Financière Européenne (40); Banques et Organisations Publiques Arabes (27.66); Banque Nationale de Paris (14.32).

40. Kuwait Pacific Finance Company Ltd.

Est.: 1975; Hong Kong; C: HK$ 25m;
Shareholders: Kuwait Investment co. S.A.K. (35); The Industrial Bank of Japan Ltd. (32); Canadian Imperial Bank of Commerce (10); Bank of New South Wales -Australia- (10); Banco de Brazil S.A.(10);
Channelling funds from Kuwait and other Middle East countries into investments in the South East Asia area.

Appendix C:

FINANCIAL BILATERAL DEALS BETWEEN OPEC AND OECD COUNTRIES[1]

(a) France-Iran

July 1974. The Iranian government agreed to deposit with the Banque de France an amount of $1 billion before end-1976. An initial deposit of $300 million was carried out in late 1974. Borrowing conditions were not disclosed.

(b) United Kingdom-Iran

July 1974. The Iranian government agreed to lend $1.2 billion to U.K. public bodies. The loan, to be drawn down over the next three years, carries a five-year maturity. Interest rates will be fixed with regard to the market situation. A first tranche was drawn by the National Water council in November ($400 million). The loan is guaranteed by the British government.

(c) Japan-Saudi Arabia

September 1974. The Saudi Arabian central bank agreed to extend a five-year loan of $1 billion in two tranches to Bank of Tokyo Ltd., the largest Japanese foreign exchange bank. The interest rate was said to be fixed at around 10½ percent.

(d) France-Iraq

December 1974. A $1 billion loan has been granted by Iraq to the French government. Details of the operation were not disclosed.

(e) Italy-Iran

December 1974. A credit line of about $1 billion has been opened by the Iranian government in favour of Italy. Details were not disclosed but the press reported that the loan would probably take the form of advance payments for investment projects to be undertaken by Italy in Iran.

[1]Information drawn from press reports. *Financial Market Trends*, OECD, Paris, February 17, 1975.

(f) France-Iran

January 1975. Iran has granted a $1 billion loan to a French public body (Commissariat à l'Energie Atomique) operating in the field of nuclear energy.

(g) Denmark-Iran

January 1975. The Iranian government has granted a loan of about $150 million to Denmark. The loan has a 5-year maturity and the interest rate is limited to the LIBO rate (1 percent spread).

Appendix D:

IDENTIFIED DIRECT INVESTMENT OPERATIONS BY OPEC COUNTRIES[1]

Country/Company	Amount (million)	Observations
Australia		
1. Clarence Oil and Minerals	A$ 10 ($11.13)	January. Acquisition of a 45% stake by unnamed Arab interests.
France		
2. Tour Manhattan	$100	April. Acquisition by Kuwait of office premises in Paris.
3. Novotel (Hotels)	—	August. Acquisition by a Swiss-based Arab Financial holding (Finial SA) of a minority stake (8.9%).
Germany		
4. Deutsche Babcock Wilcox A.G.	DM 178.3	April. Acquisition by Iran of a 25.02% stake.
5. Krupp Metalurgica Camp Limpo S.A. (Auto components)	—	July. The Iranian government plans to take a participation in the two Brazilian based Krupp ventures.
6. Krupp Industrias Mecanicas Ltd.	—	
Japan		
7. Nippon Steel Corp.	$2.5	March. Acquisition of a minority stake (0.1%) by Kuwaiti interest
Netherlands		
8. VENETA (Woven and turfed carpet manufacturer	FL 4 ($1.7)	May. Acquisition of a majority stake (52%) by a group of Arab financiers.
United Kingdom		
9. Selby Buildings	£ 50	January. Investing country Kuwait.
10. Lonrho (mining finance and industry)	£ 11.4	February. Acquisition of an 11% stake by Middle East interests.
11. Edward Bates & Sons (merchant bank)	£ 1.4 ($3.2)	May. Acquisition of a minority stake (25%) by an Arab consortium.
United States		
12. Bank of the Commonwealth (Detroit)	$10	January. Acquisition of a 32% participation by a Saudi Arabian financier.

[1]Information drawn from press reports. *Financial Market Trends,* OECD, Paris, October 1975.

International		
13. Eurodif (Nuclear energy)	$170[e]	January. Acquisition by Iran of a 10% stake (from France).
14. Intercontinental Hotels	$55	January. Acquisition of a majority stake by Iran.

[e]Estimate.

Bibliography

A. Books

Caves, Richard, and Jones, Ronald. *World Trade and Payments.* Boston: Little, Brown and Co., 1973.

Frank, H. J. *Crude Oil Prices in the Middle East.* New York: Praeger Publishers, 1966.

Fried, Edward, and Schultze, Charles, eds., *Higher Oil Prices and the World Economy.* Washington, D.C.: Brookings Institution, 1975.

Hartshorn, J. E. *Politics and World Oil Economics.* New York: Praeger Publishers, 1962.

Howe, James. *The U.S. and the Developing World: Agenda for Action 1974.* New York: Praeger Publishers, 1974.

Howe, James. *The U.S. and World Development: Agenda for Action 1975.* New York: Praeger Publishers, 1975.

Issawi, Charles. *Oil, the Middle East and the World.* La Salle, Illinois: The Library Press, 1972.

Jacoby, Neil. *Multinational Oil.* New York: Macmillan Publishing Co., 1974.

Kubbath, Abdul Amir. *OPEC: Past and Present.* Vienna: Petro-Economic Research Center, 1974.

Leeman, W.A. *The Price of Middle East Oil.* Ithaca, New York: Cornell University Press, 1962.

Machlup, Fritz. *International Payments, Debts, and Gold.* New York: Charles Scribner's Sons, 1964.

Mikdashi, Zuhayr. *The Community of Oil Exporting Countries.* London: George Allen & Unwin Ltd., 1972.

Odell, Peter. *Oil and World Power: Background to the Oil Crisis.* 3rd ed. Baltimore, Maryland: Penguin Books, 1974.

Park, Yoon S. *The Eurobond Market.* New York: Praeger Publishers, 1974.

Penrose, E. T. *The Large International Firm in Developing Countries.* London: George Allen and Unwin Ltd., 1968.

Rifai, Taki. *The Pricing of Crude Oil.* New York: Praeger Publishers, 1974.

Rouhani, Fuad. *A History of OPEC.* New York: Praeger Publishers, 1971.

Schurr, Sam., ed. *Energy, Economic Growth, and the Environment.* Resources for the Future. Baltimore: The John Hopkins University Press, 1972.

Soedersten, Bo. *A Study of Economic Growth and International Trade.* Stockholm: Almqvist & Wiksell, 1964.

Soedersten, Bo. *International Economics.* New York: Harper & Row, Publishers, 1970.

Stern, Robert. *The Balance of Payments: Theory and Economic Policy.* Chicago: Aldine Publishing Company, 1973.

B. Reports, Pamphlets and Papers

Bank for International Settlements. *Forty-Fifth Annual Report.* Basle, Switzerland, June 1975.

Bank for International Settlements. *Oil-Related Payments Imbalances.* Basle, Switzerland, March 11, 1975.

Bank for International Settlements. *The Oil Situation and the Eurocurrency Market.* Basle, Switzerland, February 1, 1974.

Beigie, Carl. *Oil Prices, Inflation, and Economic Growth.* Montreal: C. D. Howe Research Institute, September 1974.

Carroll, Joseph. "Institutional Concerns in Managing Petro-dollar Investments." Paper presented at the World Trade Institute Symposium on Investment Alternatives for the New Oil Money, New York October 23, 1974.

Dunn, Robert, Jr. "The Less Developed Countries," in Joseph Yager and Eleanor Steinberg, eds. *Energy and U.S. Foreign Policy.* Cambridge, Mass.: Ballinger Publishing Co., 1974.

Hansen, Roger. "The Emerging Challenge: Global Distribution of Income and Economic Opportunity," in James Howe, ed. *The U.S. and World Development: Agenda for Action 1974.* New York: Praeger Publishers, 1974.

Heller, H. Robert. "International Reserves and World-wide Inflation." Washington, D.C.: International Monetary Fund, July 29, 1975.

International Monetary Fund. *Annual Report 1975.* Washington, D.C., September, 1975.

International Monetary Fund. *Eurocurrency Markets.* Washington, D.C., September 20., 1974.

Investment and Property Studies, Ltd., ed. *Oil and International Money.* Transcript of a Conference held in London, April 18, 1974.

Irving Trust Company. *The Economic View from One Wall Street.* New York, March 20, 1975.

Johnson, Harry. "Towards a General Theory of the Balance of Payments," in Caves and Johnson, eds. *Readings in International Economics.* Homewood, Ill.: Richard D. Irwin, Inc., 1968.

Junz, Helen. "Some Observations on the Worldwide Intensification of Inflation," in Joint Economic Committee of the U.S. Congress, *Studies in Price Stability and Economic Growth.* Washington, D.C., August 5, 1975.

Levy, Walter. *Future OPEC Accumulation of Oil Money.* New York: W. J. Levy Consultants Corp., June 1975.

Lionel D. Edie & Company. *International Review.* New York: February 1975.

Machlup Fritz. "The Eurodollar System and its Control," in *International Monetary Problems,* Papers and proceedings of a Conference at the American Enterprise Institute for Public Policy Research, Washington, D.C., September 1967.

McNamara, Robert S. Address to the Board of Governors of the World Bank Group, Washington, D.C., September 30, 1974 (Washington, D.C.,: World Bank, 1974.)

McNamara, Robert S. Address to the Board of Governors at the World Bank Annual Meeting, Washington, D.C., September 1, 1975 (Washington, D.C.: World Bank, 1975.)

Michalopoulos, Constantine. "Financing Needs of Developing Countries: Proposals for International Action." *Essays in International Finance,* No. 110, Princeton University, June 1975.

Organization for Economic Cooperation and Development. *Flow of Resources from OPEC Members to Developing Countries.* Paris, December 6, 1974.

Paine, Webber, Jackson & Curtis, Inc. *Economics and Portfolio Strategy.* New York, November 22, 1974.

Pollack, Gerald. "Are the Oil Payments Deficits manageable?" *Essays in International Finance,* No. 111, Princeton University, June 1975.

Rybczynski, T. M. "Capital Requirements for Energy Development in the Main Industrial Countries." Paper presented to the British-North American Association, April 1974.

Schwartz, A. N. R. *The Effects of the Rise in Oil Prices on the Economy of Industrialized Countries: A Methodological and Quantitative Investigation.* The Hague: Dutch Central Planning Bureau, 1974.

Solomon, Robert. "Impact of Higher Oil Prices." Address before the Fourth European Institutional Investor Conference, London, December 3, 1974.

Stem, Carl. "The Eurocurrency System: Problems and Prospects." Paper presented to a Eurocurrency Conference at the American Enterprise Institute for Public Policy Research, Washington, D.C., October 17-18, 1974.

Triffin, Robert. "The Threat to World Economic Order: The World 'Infession' and the Oil explosion." Paper prepared for the Japan-U.S. Assembly at the American Enterprise Institute, Washington, D.C., April 7 and 8, 1975.

U.S. Treasury Department. *The Absorptive Capacity of the OPEC Countries.* Washington, D.C., September 5, 1975.

Williamson, John. "The Impact (of the Oil Price Increase) on the International Economy." Paper presented at a Brookings Institution Conference on November 18-20, 1974.

World Bank. *Energy Supply/Demand Outlook: 1980-1985.* Washington, D.C., July 18, 1974.

C. Articles in Periodicals

Akins, James. "The Oil Crisis: This Time the Wolf Is Here." *Foreign Affairs,* April 1973.

Alexander, Sidney. "Effects of a Devaluation on a Trade Balance." *IMF Staff Papers,* April 1952.

Alfonzo, Perez. "The Organization of Petroleum Exporting Countries." *Monthly Bulletin,* No. 2, Ministry of Mines & Hydrocarbons, Caracas, 1966.

Aliber, Robert. "Impending Breakdown of OPEC Cartel." *The Wall Street Journal,* March 20, 1975.

Allen, Polly Reynolds. "Can the Eurocurrency Market Finance the Oil Deficits?" *The Banker,* November 1974.

Atkin, John. "Eurodollar Interest Rates." *The Banker,* February 1973.

Atkin, John. "U.S. and Euromarket Interest Rates—A Measure of the Euromarket's Health." *Euromoney,* December 1974.

Bhagwati, J. "Immiserizing Growth: A Geometric Note." *Review of Economic Studies,* June 1958.

Clendenning, E. Wayne. "Eurodollars and Credit Creation." *International Currency Review,* March/April 1971.

Corden, W. M. "Implications of the Oil Price Rise." *Journal of World Trade Law,* March/April 1974.

Corden, W. M., and Oppenheimer, Peter. "Basic Implications of the Rise in Oil Prices." *Moorgate and Wall Street,* Autumn 1974.

Emminger, Otmar. "International Financial Markets and the Recycling of Petrodollars." *The World Today,* March 1975.

Farmanfarmaian, K., Gutowski, A., Okita, S., and Roosa, R. "How Can the World Afford OPEC Oil?" *Foreign Affairs,* January 1975.

Fieleke, Norman. "Oil and International Payments." *New England Economic Review,* November/December 1974.

Friedman, Milton. "The Eurodollar Market: Some First Principles." *The Morgan Guaranty Survey,* October 1969.

"The Growing Use of Arab Currencies." *The Banker,* March 1975.

Haberler, Gottfried. "The Market for Foreign Exchange and the Stability of the Balance of Payments: A Theoretical Analysis." *Kyklos,* Vol. III, 1949.

Hamilton, Adrian. "End of Old Assumptions on Oil Prices: Not the Oil Price Increase Itself, But the Way the Decision Was Reached, May Be Most Important for Consumers." *The Financial Times,* September 29, 1975.

Hein, John. "Oil Money and World Payments: The Billion Dollar Unsure Thing." *The Conference Board Record,* September 1974.

Hewson, John, and Sakakibara, Eisuke. "The Eurodollar Deposit Multiplier: A Portfolio Approach." *IMF Staff Papers,* July 1974.

Hewson, John, and Sakakibara, Eisuke. "The Impact of U.S. Controls on Capital Outflows on the U.S. Balance of Payments: An Exploratory Study." *IMF Staff Papers,* March 1975.

Hicks, J. R., "An Inaugural Lecture." *Oxford Economic Papers,* June 1953.

Hirsch, Fred. "The Oil-Financing Conundrum." *The World Today,* March 1975.

Houthakker, H. S., and Magee, Stephen. "Income and Price Elasticities in World Trade." *The Review of Economics and Statistics,* May 1969.

Johnson, Harry. "Economic Development and International Trade." *Nationaloekonomisk Tidsskrift,* Vols. 5-6, 1959.

Josephson, Keith. "The Eurocurrency Market." *The Journal of Commerce,* September 30, 1974.

Klopstock, Fred. "Money Creation in the Eurodollar Market—A Note on Professor Friedman's Views." *Monthly Review,* Federal Reserve Bank of New York, January 1970.

Kuczynski, Pedro-Pablo. "Recycling Petrodollars to the Third World - II." *Euromoney,* December 1974.

Latour, Pierre. "Euromarkets Wait for LDCs' Credits to be Repaid." *Euromoney,* October 1975.

Lee, Boyden. "The Eurodollar Multiplier." *The Journal of Finance,* September 1973.

Makin, John. "Identifying a Reserve Base for the Eurodollar System." *The Journal of Finance,* June 1973.

Mayer, Helmut. "Some Analytical Aspects of the Intermediation of Oil Surpluses by the Eurocurrency Market." *Banca Nazionale del Lavoro Quarterly Review,* September 1974.

Melville, Michael. "The Arcru: A Progress Report." *Euromoney,* July 1975.

Monthly Economic Letter. New York: First National City Bank, July 1974, June 1975, and August 1975.

Nelson, W. L. "Sulfur Contents of U.S. Crude Oils." *Oil and Gas Journal,* August 17, 1970.

O'Brien, Richard. "Assessing the Credit Risks of the Developing Countries." *Euromoney,* October 1975.

"Oil-Price Rise Will Not Equal 10% OPEC Set." *The Wall Street Journal,* September 30, 1975.

"Oil-Price Shaving Is Spreading in OPEC; Broad 'Cheating' by Members Seen Next." *The Wall Street Journal,* July 18, 1975.

Park, Yoon S. "Structure and Function of the Eurocredit Market." *Euromoney,* April 1974.

Park, Yoon S. "A New Philosophy for Euromarket Banks." *Euromoney,* June 1975.

Pollack, Gerald. "The Economic Consequences of the Energy Crisis." *Foreign Affairs,* April 1974.

Riboud, Jacques. "Why Rueff Is Wrong Over the Euromarkets." *Euromoney,* September 1975.

Rueff, Jacques. "How the Euromarkets Contribute to World Inflation." *Euromoney,* April 1975.

Salomon, Walter. "Bonds for the Arab Oil Funds." *The (London) Times,* January 29, 1975.

"Tax on Petrodollar Deposits?" *International Insider,* January 27, 1975.

Tumlir, Jan. "Oil Payments and Oil Debt in the World Economy." *Lloyds Bank Review,* July 1974.

Ulin, Robert. "Oil Payments and the Economy." *The Conference Board Record,* May 1975.

Wallich Henry. "Longer-run Aspects of International Monetary Reform." *Aussenwirtschaft,* September 1975.

Waverman, Leonard. "Oil and the Distribution of International Power." *International Journal,* August 1974.

Weissmuller, Alberto. "London Consortium Banks." *Journal of the Institute of Bankers,* August 1974.

World Financial Markets, New York: Morgan Guaranty Trust Company, January 21 and October 20, 1975.

"Yamani: Oil Prices Rose by Only 6.5%." *The Journal of Commerce,* October 8, 1975.

Zaki, Hassan Abbas. "Arab Financial Institutions and a New Financial Market." *Euromoney,* July 1975.